THE
SUGGESTIBLE
BRAIN

THE
SUGGESTIBLE
BRAIN

THE SCIENCE AND MAGIC
OF HOW WE MAKE UP OUR MINDS

AMIR RAZ, PHD

hachette
BOOKS
NEW YORK

Hachette Go, an imprint of Hachette Books

Hachette Book Group

1290 Avenue of the Americas

New York, NY 10104

HachetteGo.com

Facebook.com/HachetteGo

Instagram.com/HachetteGo

First Edition: October 2024

Published by Hachette Go, an imprint of Hachette Book Group, Inc. The Hachette Go name and logo is a trademark of the Hachette Book Group.

The Hachette Speakers Bureau provides a wide range of authors for speaking events. To find out more, go to hachettespeakersbureau.com or email HachetteSpeakers@hbgusa.com.

Hachette Go books may be purchased in bulk for business, educational, or promotional use. For information, please contact your local bookseller or Hachette Book Group Special Markets Department at: special.markets@hbgusa.com.

The publisher is not responsible for websites (or their content) that are not owned by the publisher.

Print book interior design by Jeff Stiefel.

Library of Congress Control Number: 2024939979

ISBNs: 9780306833434 (hardcover); 9780306833458 (ebook)

Printed in the United States of America

LSC-C

Printing 1, 2024

To my family, friends, mentors, and students

CONTENTS

CONTENTS

INTRODUCTION

A Journey from Magician to Scientist

"Conjuring is the only absolutely honest profession—
the conjuror promises to deceive, and does."
—Karl Germain, magician and lawyer

I magine a magician performing onstage in front of a large crowd in the late 1980s. He is young, an older teenager at most. Comfortable and confident in the limelight, he captivates his audience, who ooh! and ahh! and applaud after every inexplicable feat of theatrical mind reading and mental acrobatics.

But there's another sound, a distant muttering from the back. It's the low rasp of a middle-aged man, sputtering out criticisms in a run-on commentary on the show. As the show continues, he grows louder, eventually grabbing the attention of the magician. "I betcha wouldn't pull off any of that crap with *me* onstage," the man—let's call him Mr. Bling—blurts out. The magician remains unfazed and carries on with his performance.

Egged on by some members of the audience who found the developing conflict entertaining, the tension in the room became palpable. I felt it too. I was there, you see, in the thick of it all. I was the magician.

Fortunately, I had an emergency plan in place, precisely for situations of this sort. It was a plan I had never used before, but then again, I had never encountered a heckler so persistent that he could bring a vital show to its premature conclusion. I made brief eye contact with the one member of the audience who was my shill—my dutiful stooge, the ally I could summon to assist me with an alternative final act and wrap up the show before all hell broke loose. She turned pale when I gave her the prearranged signal to come to my rescue, but she quickly collected herself and zipped up her jacket to confirm that she was ready.

For a professional magician, using an accomplice—"confederate" is the technical term—is the moral equivalent of blasphemy. On the one hand, anything goes in magic: conjurers lie without hesitation, going to great lengths to facilitate an illusion or a theatrical "effect." On the other hand, resorting to the services of a confederate constitutes such bad form among practitioners of the magical arts that, when I had devised my disaster plan and planted my associate in the audience, I made sure she understood that I'd never call upon her other than as a very last resort.

"And now, a special treat," I transitioned. "Last but hardly least, I'd like to demonstrate a mystery that the greatest of wizards, let alone my young self, have yet to fully understand. It requires a volunteer who is left-handed . . . double-jointed . . ."—I paused for dramatic effect so that the audience could follow each condition—"still has all their wisdom teeth . . . is a Scorpio, and has never met me before!" From the corner of my eye, I monitored the general area of my confederate, expecting to see a raised hand for me to call on. But another sudden commotion rendered me speechless: a large figure was running down the aisle toward me.

"Perfect!" Mr. Bling shouted as he mounted the stage, my sidekick too far away to come even as a close second. The crowd grew quiet as it sensed my discomfort standing next to him. "Left-handed with a nasty hook, double-jointed, mouth full of original teeth, never seen ya pretty face before, and a mighty Scorpio with a deadly sting," Mr. Bling roared. "Let's see ya pull your brain games on me, Wonder Boy." The night was certainly not shaping up to be my best, but I went ahead and proceeded to perform the trick the way I usually would.

By some wizardly stroke of magic, however, when Mr. Bling got to the climax of the effect, instead of boisterously bashing me or the trick, he became limp and wholeheartedly amazed. Such a wave of emotion came over him that, for a few seconds, it seemed that he was weeping on stage. The sight of a burly, loud-mouthed heckler, with eyes and mouth wide open, transforming into an awestruck lump of sentimental mush was not lost on the audience. After all, how many performers do you know who could tell you some known-only-to-you bits of intimate information? I revealed to Mr. Bling what his mother, on her death-bed, moments before she passed, had whispered in his ear. His emotional response was palpable. Moreover, this presentation had moved my confederate (and the audience) in a profound way, as it should have. Neither was the marvel lost on me; I must have looked baffled. It was a tremendous ending.

The applause was thunderous and carried on well after I returned backstage, where my confederate was waiting on me. She grabbed me by the shoulders and looked like she had just witnessed a miracle: "How did you manage it? I don't understand," she whispered with her eyebrows curving above her soft eyes.

"I'm not entirely sure," I said after a long pause. "I had this fluky intuition," I added quietly, "a kind of premonition, really."

Again, she looked at me with a glazed stare. "But you did it. Without me. On your own. That's insane, absolutely incredible!"

After that show, and by her own admission, we both decided that I didn't really need a confederate. What's more, for years after, she sang my praises, saying that I could do tricks that other magicians could not. When I was nearby, I'd blush and ask her to stop.

"But it's true," she'd insist. "You really have a special gift."

"Don't make a fuss; perhaps I just got lucky, or maybe there's a different explanation," I'd protest weakly, but she would ignore my bashfulness and wax eloquent, to anyone willing to listen, about what had transpired that day.

* * *

Now let's take a step back and walk through what happened here. We all know that "magicians *never* reveal their secrets." However, sometimes unraveling tricks divulges their genius, and I share this story because it illustrates how to work with suggestibility. Because I am a cognitive brain scientist, explanations in this book lean heavily on mental action and the neurosciences. In science, as in magic, things usually happen by design—serious scientists and magicians mull over, plan, rehearse, and polish before they're ready to roll out. This show was no exception.

Such an honest liar I was that all my shows opened with an explicit disclaimer: "I stand before you wearing the hat of a showman as I perform theatrical mind reading for your entertainment pleasure. The tricks and illusions you are about to witness, loosely referred to as 'mentalism,' are all part of the magical arts." However, my prompt rarely worked. After most shows, a line of people would often form to consult with me about some pressing life dilemmas—from treatment for a dying aunt to investing in the stock market—as if my acting performance implied that I possessed real omniscient ability. Like others before me, I discovered that disclaimers don't always work as intended.

Once, when a woman, whom I vaguely recognized as one of my

mother's acquaintances, cornered me for advice after a show, I explained to her at length that it was all a performance and that I don't really have any special powers to read minds. She smiled at me knowingly and whispered in my ear: "You don't have to be so modest. I know your family... You are too young to realize how gifted you are and what powers you have!"

What can I say? Performing as a mentalist has undoubtedly taught me a great deal about human behavior, including something that many performers and scholars had discovered long before me: given the right context and suggestion, people don't always listen to and understand what you say. Instead, they hear and see what *they think* is happening.

Whereas a moving theatrical performance may propel children to believe magic is real, for most adults—including Mr. Bling and my confederate—it's unlikely that any effect would go quite *that* far. Yet the right suggestion can tickle people, even skeptics, so that they may move closer to considering an unlikely possibility. In other words, a dramatic experience can increase suggestibility and communicate a specific underlying suggestion.

Consider my confederate: not only was she my trusty assistant, she was familiar with the show and had seen variations of it many times before this specific performance. With my encouragement, she must have viewed herself as an essential safeguard against a show going wrong, an insurance policy against a catastrophe. But after I was able to pull off the trick even with a harsh and unfriendly disaster of a participant, she also started rethinking the necessity of her role. This is the power of suggestion: she came to this conclusion on her own, without my explicit guidance. Moreover, she started entertaining the possibility that "this 'Wonder Boy' really has a special gift." That's a more nuanced suggestion.

Although it's been nearly four decades now, I've never told her that

Mr. Bling was my uber-plant, my meta-stooge. Yes, I had devised the whole thing well in advance. Magicians are sneaky actors, and Mr. Bling—who has since passed on—was a fellow magician, mentor, friend, and an excellent dramaturge. We had discussed our deceptive plot and rehearsed it for a while. Having begun to feel uncomfortable about carrying around a "helper," I wanted to let my shill go in a magical way, leaving her bedazzled while giving me good public relations. In this case, I was able to achieve both.

* * *

When I was a young magician, my friends and I would put on stage hypnosis shows. We'd make participants cluck like chickens, act out James Bond "007" maneuvers, and forget their own names. Tears would roll down spectators' cheeks, the audience doubled over in laughter and amazement. Back then, I was hardly the scholar and cognitive neuroscientist that I am today. My buddies and I got all our hypnosis know-how from a popular booklet we bought at our local magic shop. For the life of me, I couldn't fathom why fully grown people would spend their hard-earned cash to voluntarily humiliate themselves on a stage commanded by teenagers.

A few degrees—undergrad in computer science, master's in computational neuroscience, plus a PhD in cognitive psychology and information processing in the brain—and decades later, the scientific study of suggestion has become a major part of my life. Today, I am one of the leading experts on what suggestion can and cannot do, as well as the neural correlates of hypnosis, motivation, and conviction. What I discovered was that "mindset" was not just an ethereal, abstract notion, but also a brain state that can be characterized by neurophysiology, functional neuroanatomy, and behavioral outcomes. I can now explain not only how suggestion can affect our blood flow, body temperature, and motor tics, but also how suggestive information actually shapes our brains.

After thirty-five years of study and research as a neuroscientist and professor of psychology, my findings are more relevant today than ever. We've all watched the gradual erosion of science and truth in modern culture. I also learned that the way professionals across industries—including negotiators, interrogators, lawyers, marketers, influencers, and others—use suggestion is very different from the way scientists, researchers, and scholars study it.

Truth is, the power of suggestion lies in the gap between our interactions and our perceptions of others, and that has implications for society. Against a backdrop of political divisiveness and generational differences, this book speaks to critical issues in our country and addresses everyday situations where readers may use suggestion to their benefit, or avoid influence from unwanted suggestions. My intention in writing *The Suggestible Brain* is to harness the science of suggestion to propel change, to protect against manipulative misinformation, and to better regulate our internal mental worlds.

Here, I share some of my best stories about the science of suggestion, the bread and butter of my professional life. Throughout my academic career, I have focused specifically on the role that suggestion plays in the effectiveness of psychiatric medications, alongside how attention, self-regulation, placebo phenomena, and expectations affect human consciousness. For example, sometimes placebos work even when people know that they are taking them. The fields of developmental psychopathology and transcultural psychiatry teach us how much our communities and social standards can shape our thoughts, emotions, and cognitive processing. And using imaging of the living human brain and other state-of-the-art techniques, my research has helped unravel the mental experiences of hypnosis and meditation.

The bottom line is simple: everyone is suggestible. Whether we realize it or not, suggestions hack our minds and shape our realities every day. By standing on the bridge between illusion and science, we can learn how to transition away from the realm of manipulation into taking better charge of our own subjectivity.

Suggestion can make cheap wine taste like Château Margaux, warp our perception of time, and alter our memories. We know that psychedelic substances, illusions, and other dramatic effects can also alter our levels of suggestibility. But, paradoxically, being susceptible to suggestions hardly means we're feeble-minded or gullible. In this book, we look at why, as well as answer scientific questions like:

- What's the evolutionary advantage of being suggestible? How did humans, compared to other preexisting species, become especially good at being susceptible to suggestion and to adopting different mindsets?
- In what ways can these mental states influence our decisions and choices across life?
- What are the limits of suggestion?
- Can we become more or less suggestible, and what are the specific domains wherein suggestions can make the greatest impact?
- Can suggestions and mindsets effectively treat mental health conditions, including depression and anxiety? What are the underlying brain mechanisms that afford such possibilities?
- Why are we more likely to believe information that already aligns with our beliefs?
- Why do placebos work even when the people who take them know that they are taking inactive pills? And why are people who are "highly susceptible to suggestions" not the same as "good placebo responders"?
- How can suggestion lead entire groups to believe and act

in line with a specific agenda, as in the seventeenth-century Salem witch trials?

- Why do some election campaigns use "push poll" tactics to influence our opinions and voting behavior? Can we learn how to effectively thwart or successfully resist such premeditated assaults on our minds?
- How do people weaponize suggestion in the form of gaslighting and mental abuse?
- How do mass suggestion and social contagion work together to influence behavior?
- How can suggestions help fight racism, bias, and bigotry? Can they backfire and create the opposite effect?

Ordinary people, like you and me, can manifest magical effects given the right suggestion, encounter, timing, and delivery.[1] People exposed to phony "poison ivy"—an otherwise innocuous plant that they think will give them an allergic reaction through contact with its oily resin—can develop real rashes,[2] and unsuspecting folks drinking "regular" coffee, although it's actually decaf, can still show the increased motor performance and elevated heart rate associated with genuine caffeine.[3]

At the same time, suggestibility provides an important set of learning tools to navigate the world. Certain suggestions can help us lose weight by drinking milkshakes,[4] manage our symptoms with careful language and nutritional labels,[5] shape our food choices by recasting memory traces,[6] and even help us reap the same benefits of regular exercise through housekeeping chores[7] and rigged step counters.[8] And suggestion even appears to improve our vision,[9] regulate our body temperature,[10] control our digestive system,[11] and reduce our experience of pain.[12] These outcomes transcend what people often collectively refer to as "placebo effects"—they form real instances when expectations shape physiology.

Understanding *how* the suggestible mind operates, as well as the limits of suggestion, reveals a great deal about human nature, what makes us tick, and the relationship between what we expect and what actually happens. Moreover, it's applicable—for example, how do we deal with suggestive interrogations and uncertain memories; how do we improve motor tics and other uncontrollable behaviors; and how do we use suggestion in the treatment of depression and anxiety, group settings, social injustices, and more?

Above all, the science of suggestion reveals the fragility of truth itself. Our reality lies in the hands of savvy advertisers and clinicians, in the words of charismatic leaders and politicians, under the spell of social media and the algorithmically crafted narratives of digital platforms. Armed with the information herein, you will embrace and benefit from empowering suggestions, identify and avoid the pitfalls of negative ones, and successfully navigate both your internal and external worlds. Whether we move away from astrology and closer to astrophysics or hedge our faith on conspiracy theories and televangelists, suggestion may well make the difference.

THE
SUGGESTIBLE
BRAIN

CHAPTER 1

The Magic of Suggestion

As a young performer, I started out with simple suggestions. I'd say, "Kentuckians pronounce the capital of Kentucky as neither Looey-ville nor Lewis-ville. Do you know how they say it?" (The capital of Kentucky, in case you don't know, is Frankfort.)

By middle school, I refined my ways and took to more subtle "adult" suggestions. For example, I'd take five cards out of a deck, arrange them casually in front of a spectator, and say the following *[while simultaneously motioning]*:

"Look at all these five cards carefully: we have one picture card *[pointing to the queen in the middle]*, one ace of diamonds *[gesturing to the ace]*, and one black card *[nodding to the seven of clubs; the other two cards were the four of hearts and the five of diamonds]*. In a moment, I will ask you to select one of them. Don't do it just yet because your choice should be completely free and entirely uninfluenced by me. Remember, whatever card you choose out of the five, make sure I have nothing to do with it. Okay? Good, now go ahead and select your card..."

I'd pause for a few short seconds, letting everything sink in, and then say:

"What card did you think of?"

*   *   *

It's amazing how different people consistently and reliably fall for this suggestion. I wonder whether you, too, chose the four of hearts. Whereas most people get Frankfort rather quickly, they largely remain in the dark about the four of hearts.

This trick uses a technique that I call "mental contamination." By mentioning and pointing out the queen, ace of diamonds, and black seven, I am reducing the chances you'd choose them. After all, I am asking you to select a card free of my influence. By extension, I am also contaminating the five of diamonds because I repeatedly say "five" and because I have contaminated the diamonds suit by mentioning the ace of diamonds. The only "pristine" choice that remains, then, becomes the four of hearts. It's the only nonblack, even-numbered card; spatially, it sits to the unscathed left of the queen (I have contaminated the right-hand side), and it bears a red suit that I haven't yet spoken of—hearts. All these little elements increase the likelihood that people will home in on the four of hearts.

Having this acted out by a confident, expert performer who can navigate and direct the situation with the right timing and gesticulations makes this outcome even more probable. It creates a feeling of free choice and complete autonomy that the spectators—you included—purport to have.

Whether or not I've successfully duped you into settling on this card—it's actually harder to effectively transfer this suggestion via text than through an in-person performance—you should appreciate that suggestion is something not reserved only for magicians. Instead, it's a learnable technique that anyone can master. Even a person with a small amount of talent and dedicated diligence can become a decent chess player after a few years, albeit not a Bobby Fischer. The same is true for playing a musical instrument, public speaking, and cooking.

*   *   *

In my teens, I ran into a fascinating "little red booklet" on hypnosis. Inspired by the dictates of its twelve pages, and without specific or extensive training in hypnosis, psychology, or mental health, my friends and I came up with effective acts, almost entirely hinging on a bold display of showmanship, charisma, and suggestion. Without flinching, some of my friends took their performances to town. I was genuinely amazed by the success of these shows because we knew very little, if anything, about how hypnosis worked. *Is it all role-play?* I remember thinking, *Why are volunteers following our ludicrous, "pseudohypnotic" suggestions?*

At some point, I considered whether the people who volunteered to come on stage were, perhaps, not the sharpest knives in the drawer. It seemed unlikely because some of them were highly educated and prominent members of their communities, but it wasn't until 1985, when I read about the hypnotic experience of Richard Phillips Feynman—winner of the 1965 Nobel Prize in Physics and one of the greatest scientific minds of the twentieth century—that I decided to look into hypnotic suggestion more scientifically.

Feynman shared his experience of volunteering for real hypnosis demonstrations on three separate occasions. During one relatively private forum at Princeton University, Feynman described how he was slightly "fogged out" by the hypnotic induction ritual and suggestion—for example, that he couldn't open his eyes—but that he was pretty confident that he could open them and just kept them shut to play along so as not to embarrass the guest gentleman, a professor of psychology who was the hypnotic operator. During this preliminary session, the psych prof determined that Feynman and a few of his colleagues were sufficiently hypnotizable to participate in a demonstration in front of a large, live audience.

A week later, Feynman and his select peers went through these motions again as part of a public demonstration before the whole Princeton graduate college. This time the effect was stronger, and the psychology prof gave Feynman all sorts of suggestions to do things that he couldn't normally do. At the end, the hypnotist said that after Feynman came out of hypnosis, he'd not return to his seat directly—the natural way to go—but would walk all the way around the room and go to his seat from the back. Feynman, who had cooperated with all the suggestions up to that point, decided to resist and disobey this last one. When it was time to get up and walk offstage, he started making his way straight to his seat. But then an annoying feeling came over him; he felt so uncomfortable that he couldn't continue. Instead, he walked all the way around the hall . . .

In a later, third encounter with hypnosis, a woman hypnotist told Feynman that she would light a match, blow it out, and immediately touch it to the back of his hand. She assured him that he would feel no pain. Feynman was skeptical but agreed to go along with this hypnotic suggestion keeping his eyes closed throughout. When the woman took a match, lit it, and touched it to the back of his hand, Feynman felt only slight warmth. With his eyes still closed, Feynman explained this to himself: *She probably lit one match and touched another to the back of my hand*. But when he got out of hypnosis, opened his eyes, and looked at the back of his hand, he was surprised to find a burn right there. Soon a blister appeared and grew. But it never hurt at all, not even when it broke.

* * *

Feynman was one of my intellectual idols, but I was still uneasy about how my magician friends were merely playacting the role of hypnotists; they were not even trying to do any traditional magic. Their show delivered a charade, a theatrical display of hypnosis, and people were falling for it en masse.

One day I happened upon a stage hypnosis show by a Russian master. He was a prime example of a "hypnotic operator": smooth medicinal appearance, complete with stethoscope and starched white coat, bushy eyebrows riding over penetrating eyes, silver hair pulled back into a ponytail, trimmed goatee under thinly pursed lips, and a dramatic inflection alongside an impressive baritone. The only problem with this formidable package was that he spoke only Russian. So the organizers hired a translator—a soft-spoken man with a velvety, radiophonic voice—to echo the hypnotic Russian utterances in the local parlance so that people could understand and follow. What transpired that evening taught me a valuable lesson about human suggestibility.

The master took center stage. He was basking in the spotlight, articulating affected Russian sentences accompanied by large hand movements and heightened facial expressions. After a short pause, the radio announcer would chime in from a little podium on the far-right side of the stage. The podium covered most of his skinny appearance, leaving only his upper body visible. Yet when he spoke, everyone would move their gaze to look at his talking head. After all, we could understand what he was saying. The show progressed: a bit of foreign mumbo jumbo adorned with melodramatic gesticulations from center stage followed by a cheerful translation we could relate to and comprehend.

Participants came onstage and got hypnotized, all by following translated suggestions from the radio telecaster. After about forty minutes of this ping-pong dynamic, just before the break, it became apparent that the Russian master was getting in the way. People were interested in what the gentle radio presenter had to say, and they started to feel that the Russian master was disrupting the show with his exaggerated shenanigans.

During intermission, I ran backstage. I had performed in that venue previously and knew how to get from the foyer to the dressing rooms. I found the radio announcer sipping from a glass of cold water and wiping down his sweaty forehead with a white handkerchief.

"The show is going really well," I greeted him without introducing myself. "Have you been doing hypnosis for a long time?"

Mr. Announcer looked at me warily and lowered his voice in a heartening reveal: "I have no idea what's happening. I know nothing about hypnosis and have never done anything of this sort before. They asked me to translate simultaneously, and that's what I am doing, but I feel things are getting a bit out of control. It almost feels like *I* am the one hypnotizing the people."

Like him, I was shocked that a man with no training in hypnosis could mesmerize an entire audience just by translating utterances. He was merely a talking head—certainly not a hypnotist, just a conduit to the master. But then I thought about my adventurous magician friends, and the whole thing came into focus. It's not about what you know or even what you say—it's about what they *think* you know and what they *think* you say.

Hypnotic suggestion is especially interesting: when you say to yourself, "I can do that, but I won't," that's another way of saying that you can't do it. Hypnosis seems to violate our assumptions about the voluntary control of behavior. We typically assume that mature adults exert voluntary choice in their actions. When hypnotic suggestions subvert voluntary control—as in the examples from Feynman—we feel surprised, disturbed, or both. But psychological analysis and experimental evidence propose that our common assumptions about voluntary control of the mind and behavior may be a bit naïve.

In 2005, when I was a professor at Columbia University and the New York State Psychiatric Institute, the *New York Times* sent a reporter and photographer to interview me for a story they were running about my research on hypnotic suggestion. The photographer introduced himself briefly and asked permission to roam about and look for interesting visuals. The reporter and I sat at a table and talked while the photographer walked around the lab snapping photos from different angles, juggling his multiple cameras, lenses, and lights. At some point, the reporter asked to see a demonstration of how I administer hypnotic suggestions to a participant. I called one of my volunteers into the room so I could perform a dramatization of the procedure. The three of us—reporter, volunteer, and I—huddled around a table, and I started to go through the motions of a hypnotic induction coupled with suggestion. Within a few minutes, we noticed an eerie silence: the photographer had stopped clicking photos and was slumped on a chair in the corner, deeply hypnotized![1]

In hypnosis, participants who are highly susceptible to hypnotic suggestion report and display a wide variety of fascinating responses.[2] For example, they report hallucinations (hearing or seeing something that's absent), negative hallucinations (not smelling or feeling something that's present), and pain control; they can display selective amnesia (inability to recall information), changes in mood, and even partial paralysis. Because these effects appear theatrical, it's natural to associate them with dramatic causes.[3] Whereas researchers agree that many of these reports and responses reflect genuine changes in experience,[4] some scholars attribute them to social dynamics, advancing different theories about how these effects come about—for example, as a function of compliance with perceived expectations and the demand characteristics of a situation.[5]

The history of suggestibility intimately connects with that of hypnosis.[6] An early form of psychotherapy, hypnosis has been tarnished by a checkered history: stage shows, movies, and cartoons that perpetuate specious myths, and charlatans who unabashedly offer hypnotic enlargement of anatomical organs and other medical hyperbole.[7] Even today, scientists are still unraveling how hypnosis works. But little is mystical about this potent technique, which capitalizes on the power of suggestion and serves as a helpful tool for both clinicians and researchers.[8] We have developed a few critical insights concerning suggestion and suggestibility: we now know what it can do—for example, help with flight anxiety or smoke cessation—and what it cannot do—grow a new arm or straighten misaligned teeth—and for whom.

Hypnosis research has increased markedly in both quality and quantity since the introduction of standardized, reliable measures for assessing response to hypnotic suggestion. Some of these tools were firmed up a few decades ago at Harvard[9] and Stanford[10] using terminology that draws on a variety of close words: hypnotizability, suggestibility, hypnotic responsivity, hypnotic susceptibility, and hypnotic suggestibility. These words largely mean the same thing, and in the interest of simplicity, I will avoid their nuances and use them interchangeably.[11] The bottom line is that by using a number of psychological tools, mostly interview-like questionnaires, today we can quickly (in about an hour) and effectively (with reasonable specificity and accuracy) determine how hypnotizable people are—for example, by placing them on a scale from low to high.

Moreover, we have developed a few critical insights concerning human suggestibility. For example, whereas hypnotizability refers to the increase in suggestibility by hypnosis, in practice, hypnotic susceptibility scales can index measures of everyday suggestibility, not just of hypnotic suggestion. In other words, the academic discussion of suggestibility has transformed into, and can now directly apply to, practical, common situations. Today we refer to hypnotizability as a measure of a particular type of suggestibility—imaginative suggestibility.[12] However, many types of suggestion exist, and hypnotic suggestion is but one type.[13] Throughout the following chapters, we will explore a few different forms of suggestibility and their relevance to constructing our reality.

*   *   *

Sometime in 2007, I left Manhattan to become the Canada Research Chair in the Department of Psychiatry at McGill University in Montréal within the French-speaking province of Québec, Canada. After a long day of moving box after box into my new office at the Jewish General Hospital, I sat back in my newly minted Canadian chair to catch my breath and survey the unfamiliar landscape that was visible outside my window. It was beginning to get dark out, and the daily bustle around the building was thinning out. Against this backdrop, a faint knock grabbed my attention. The opening door revealed a thin woman in her late twenties. "Hello," she said in a tremulous voice without introducing herself, "I heard you were in town, and I was wondering whether I could speak with you briefly."

"Of course," I gestured to the chair across from me, "please excuse the mess. I literally just moved in." She sat down, visibly nervous, her big blue eyes exploring the disarray around the room. "Please tell me how I may help you," I reassured her.

"It's about hypnosis," she began. "I have read some of your papers and have seen how you hypnotized people to feel hot in subfreezing temperatures and then forget the number four and count eleven fingers on their hands."

"That's great," I nodded to her encouragingly.

"I was wondering..." she paused.

"Yes," I said after a few seconds of silence, "I am listening..."

"I was wondering whether..."

At this point, she went into some form of a lull again. Her eyes remained open as she gazed out the window, apparently daydreaming. So we sat in silence for a bit, enjoying the sunset from the window. It was my first Montréal sunset in the new office.

"Whenever you're ready, just let me know..." I gently tried to interrupt her trance-like lapse. I had to repeat myself a few times, changing cadence ever so slightly and waving my hands slowly through her line of sight. Then I just sat there quietly, waiting for her to come around. After a minute or two, she did.

"Wow, that was intense!" The young woman smiled and relaxed in her chair.

"What was?" I asked, watching her orient herself to the room.

"The session!" She exclaimed in a rapid response infused with a wave of renewed energy. "And I can see things much more clearly now. Thank you so much, Dr. Raz!" She stood up to leave, and I couldn't but marvel at the transformed person who was leaving my office. I wanted to ask about her experience and what had just happened, but she was already out the door.

"You're welcome?" I said to the empty room, trying to exhume from memory the last time I encountered a similar case.

This occurrence was neither the first nor the last time that I witnessed a highly suggestible person slip into spontaneous hypnosis. I didn't have to say or do anything special; responding to the presence of an expert, expectations took over and created a realistic experience. Sometimes the proximity of an authority figure—a doctor, a celebrity, even a stage hypnotist, shaman, or member of clergy—is sufficient to potentiate or unleash some experiences. Our expectations can sometimes create a strong subjective reality.

Shockingly, the same is true for all people, even those who are less suggestible. It's just a matter of finding the right suggestion at the right time and setting. For example, my students and I recently showed that giving study participants placebos when they think they are receiving psychedelic drugs[14] can still provide a meaningful mind-altering session.[15] More on that later.

How can "nothing" bring about such strong outcomes? The answer starts with the fact that expectations are far from being nothing. Our expectancies form very strong vectors in our responses and what we experience. Together with context, mindset, and symbolism, what we think and presume can sometimes take over what is actually happening. This "force field" is the domain of suggestion and suggestibility.

What's the Difference Between Suggestion and Suggestibility?

Most people harbor an intuitive notion of what suggestion and suggestibility mean: suggestion serves as a type of influential communication, whereas suggestibility indexes the degree to which a person will accept a suggestion and perhaps act on it. We all know that receiving a suggestion doesn't necessarily translate into a reason to do anything. But getting a suggestion may be reason enough to think about something as it draws the suggested information to our attention.

The practice of suggestion embraces the idea that our actions may improve by gleaning the experience and insights of others and that their input is worth taking into consideration. Yet humans are notoriously stubborn when it comes to accepting nudges. Some people will disregard expert advice even when they know little about a specific topic. And so, those who are less suggestible may ignore guidance on how to make a decision or conduct themselves. For them, a suggestion may be more about the overarching thoughts it triggers than about its specific message. For example, my dad, who knows little about medicine and is arguably one of the least suggestible people I know, doesn't believe in doctors. Before he needed to make a health decision, I encouraged him to consult a few top clinicians. He was skeptical of their recommendations, of course, but the process of comparing and contrasting the different opinions they presented helped him come up with his own narrative of what he should do, which was not grounded in their medical advice per se but more in a combination of associations that he came up with as a result of their accounts.

Many people prefer to ignore suggestions and favor their own perspectives. Perhaps this is a result of our obsession with winning arguments over thinking straight. However, this form of discounting acts as an anchor and prevents us from accepting different views. That's why a compromise between our cherished viewpoint and what an advice-giver is suggesting can sometimes improve our performance. For example, when my dad agreed to consult with some medical experts, it was after he had already formed his own largely uninformed opinion. At the end of the process, his final decision was some kind of a compromise between his initial position and a potpourri of the advice he heard from his doctors. Moreover, even conflicting advice— albeit a likely source of frustration—can be helpful: it may nibble at our confidence and cause us to consider our decisions more judiciously.

Humans tend to be overconfident in their decisions; overconfidence leads to underthinking.

Throughout life, some periods seem more amenable to suggestion. For example, family and friends may provide coaching about a career path or educational trajectory at crucial points in our personal and professional development; parental advice pops up as we get ready to start a family; interview strategies or salary-negotiation tricks come up when we look for a new job; and we get dating tips after a breakup, divorce, or loss of a partner. These suggestion-heavy periods allow us to receive input from others about how to effectively navigate, or simply survive, the vicissitudes of life. Sometimes we just need a jolt—its actual direction less important—when we face a life-changing event.

But the merit of suggestion hardly rests on just doing what others say. Rather, suggestions can be a tool for self-discovery and personal growth—a mirror to reflect on and perhaps question our beliefs, blind spots, and intuitions. Information from others forces us to mull things over with greater attention to detail. To ascertain whether the suggestion represents something we want, we must consider how it gels with our own aims and mores. As such, we must first ponder and clarify those goals and values for ourselves—an essential cognitive process.

* * *

Many people incorrectly confuse suggestibility with feeble-mindedness, gullibility, persuadability, or spinelessness.[16] They think suggestibility means that someone can be easily influenced or manipulated. However, rather than a measure or quality of being easily duped or cheated, suggestibility describes the ability to follow a suggestion. For example, one of my office colleagues at Cornell would become very drowsy and sleepy if I'd so much as intimate that she looked tired. An intelligent and accomplished woman, her response was "I can't help it. Once I hear a suggestion, things just happen to me."

The problem is that the terms "suggestion" and "suggestibility" are not monolithic concepts; they come in different shapes and flavors. Highly hypnotizable folks make up about 10–15 percent of the population, inclined to accept and act on hypnotic suggestions made by others. However, as you will see in the next chapter, hypnotic suggestibility tells us something about suggestibility outside of hypnosis. Children, teenagers, and adolescents tend to be more stubborn but also more suggestible,[17] perhaps as part of an evolutionary mechanism to help them learn and find their way in a complex world.[18] It's not always possible to tell whether a child will become a highly suggestible adult;[19] but we do know that following a peak around the early teenage years, hypnotic suggestibility typically drops and then plateaus into adulthood.[20]

A few short decades ago, scientists thought that suggestibility correlated with a tendency toward inordinate imagination or sensitivity. Accordingly, back then scholars compared suggestibility with being a "Trekkie," crying at the opera, falling in love with film stars, or getting so deeply engrossed in reading a book that you lose track of time or miss the phone ringing. They believed these behaviors correlated with a specific kind of emotional openness to suggestion.[21] Today, researchers construe suggestibility not just through its correlates with particular personality traits—such as absorption and dissociation—but also through the lens of neurobiological features, including neural wiring, brain connectivity, and the higher cognitive functions of our nervous system. Altogether, I collectively refer to the psychological and social science of how suggestions shape our behaviors, alongside the neural dynamic involved, as the realm of the suggestible brain.[22]

Scientists are still relatively early in fully explaining "What Correlates with Suggestibility?"[23] One idea is that suggestibility has to do with processing expectations—our brain is constantly trying to predict what will happen.[24] To make sense of the present, we try to foretell what will then transpire;[25] both suggestibility and expectancy play into this aptitude.[26]

The bottom line is that exact definitions of suggestion and suggestibility—distinct yet related—remain fuzzy, reflecting the difficulty of pinning down this pervasive yet perplexing aspect of human behavior. We generally agree that whereas hypnotic or imaginative suggestion represents the process by which a thought or idea brings about a physical or mental state, suggestibility describes the individual differences in response to suggestion.[27]

How Do We Measure Suggestibility?

Let's say you wanted to find out your intelligence quotient (IQ), rather than your suggestibility score. That's easy: your IQ score would be a number derived from a set of standardized tests designed to assess human intelligence—an IQ test.[28] Most people likely wish for their IQ score to be high enough to demonstrate that they are extremely intelligent; after all, IQ tests provide estimates of our intellectual crown jewel, even with the obvious caveat that a concrete measure of human intelligence is challenging to obtain and that there are also different types of intelligence.[29]

Researchers correlate IQ scores with different things. For example, we know that IQ impacts the likelihood of divorce. Couples with below-average intelligence are 50 percent more likely to divorce than those of above-average intelligence.[30] However, intelligence hardly predicts financial well-being; you don't have to be "smart" to be wealthy. Mensa aside, it's difficult to say what exactly IQ tests measure, but those with a high score usually see positive outcomes in education and social competence.[31] In two words, having a high IQ score, for the most part, seems like a *good thing*.[32]

How about suggestibility? Finding out how suggestible you are is not the same as finding your IQ score. It turns out we can answer this question in several ways, some more scientific than others. For example, we have psychological surveys, structured mini-interviews of sorts, which allow a trained evaluator to determine a score that places your response to suggestions on a consistent scale. Although such suggestibility tests serve as the main form of measurement and you can easily look them up online, when administered by an expert, they provide results that are extremely reliable. The psychometrics—the science behind psychological testing[33]—of these suggestibility scales is as good as, if not better than, that of IQ tests.[34]

But unlike with IQ tests, many people who take a suggestibility assessment may covertly wish that their score ends up low rather than high. Why? Because at some fundamental level, suggestibility feels like a weakness or liability.[35] It seems to flag an inability to hold your own ground, a proneness to manipulation or fantasy, and it perhaps even labels the suggestible person as easy prey for predatory types and con artists. Being highly suggestible, moreover, may reek of being naïve, dull, uncritical, or psychopathological. The fabric of reality, however, makes for a richer, more intricate garment.

When I first started assessing suggestibility, I almost exclusively used the research interviews developed at Harvard to test groups of people and at Stanford to test individuals. These tests—"structured interviews" would probably be a better name—have good agreement between them;[36] used in concert, they serve as the academic gold standard for determining suggestibility.[37] The main problem is that they are time-consuming, over an hour each, rendering them impractical for busy people to sit through. As a result, shorter suggestibility measures have sprouted, mostly driven by clinicians who need to determine the suggestibility of their patients during brief encounters. Such tests are certainly more expedient than the Harvard and Stanford scales, but their reliability sometimes leaves something to be desired.[38]

One evening in Manhattan, for example, two prominent psychiatrists who were visiting our neck of the woods invited my wife and me to join them for dinner at a nice restaurant. We had a lovely evening together, and when dessert came around, we began sharing hypnosis stories. I told them that we used hypnosis when my wife gave birth to our children. Intrigued, both hosts immediately turned to her and asked about the experience. They wanted to know how hypnotizable she was and offered to put her through a rapid clinical suggestibility test, the Hypnotic Induction Profile (HIP), that practitioners sometimes administer in about ten minutes.[39] One of them took charge and asked my wife to hold her head still, looking straight forward. Then, while she kept her head in that position, the other asked her to look upward toward her eyebrows and then toward the top of her head. While she was continuing to look upward, they both asked that she close her eyelids slowly. The amount of sclera—the white part of the eye—visible between the lower eyelid and the lower edge of the iris was the criterion they were measuring.[40] They both chimed in at the same time: on a scale of one to five (from least to most suggestible), one gave her a two and the other a five. It was an educational moment.

"That's a bit of a discrepancy, don't you think?" I pointed out to them. In science, you see, the *reliability* (or consistency) and *validity* (or accuracy) of a measure determine how well the method captures the phenomenon of interest.

The eye roll test is both unusual and controversial. Mostly, however, it's a *supposed* indicator of hypnotizability.[41] To say that we currently have either a good theory or a scientific understanding of why the amount of sclera showing between the bottom of the iris and the lower eyelid should index hypnotizability would be an overstatement.[42] Moreover, whereas the psychometrics of the Harvard-Stanford combo is admirable, most researchers find no relationship between the eye roll sign and other measures of hypnotizability.[43] And yet, some HIP-minded clinicians continue to claim that, in addition to minimal obtrusiveness, the eye roll sign is helpful in clinical settings.[44] In other words, despite well-deserved efforts to come up with novel or speedy methods to assess hypnotizability, not all ways of measuring suggestibility are reliable.

Responsiveness to suggestion seems stable over the decades of our lives, almost like a biometric fingerprint pattern; in other words, some researchers liken suggestibility to the constancy of our fingerprints, a unique physical feature that doesn't change throughout life.[45] Even in the face of such stability, however, it's been notoriously difficult to identify the correlates of suggestibility. This lacuna is disappointing because we would like to understand the underlying mechanisms. Moreover, our theoretical accounts propose various predictors of hypnotic suggestibility that should be detectable, yet we are still looking for them.[46] So, the research scientist and the treating clinician may come up with different suggestibility scores for the same person.[47]

* * *

In the early 2000s, I traveled around Asia for several invited talks. One of my first stops was Singapore, where I would give a lecture on the topic of hypnosis and suggestibility at a local university and make time for questions at the end. But despite a large lecture hall full of sparkly, curious eyes, I was surprised by the lack of raised hands. My host, however, did speak up and informed me that many students wanted to learn more about the Harvard and Stanford scales, so I offered to hold a demo. Testing them as a group on the Harvard scale, I happened upon a shocking discovery: nearly everyone in attendance was highly suggestible! Even when I followed up with some of them individually using the Stanford scale, I was further stunned by how suggestible most students appeared.

This experience was particularly striking because just days before my arrival in Singapore, I held a similar workshop at the University of California in Los Angeles (UCLA), where the outcome was quite different. The UCLA demo was a textbook example, with only about one in a dozen people receiving the appellation "highly suggestible." In all my decades of experience working with suggestibility scales, I had never seen such a large number of "highly suggestibles" as I had in Singapore. *How can this be?* I thought to myself.

Suggestibility scales usually encourage participants to imagine certain things. For example, one item from the Harvard Group Scale of Hypnotic Susceptibility asks that you imagine a buzzing fly, which becomes so annoying that you need to shoo it away. Afterward, you fill out a form, which asks, "Would you estimate that *an onlooker* would have observed you make any grimacing, any movement, any outward acknowledgment of an effect (regardless of what it was like subjectively)? Circle one:

A. I did make some outward acknowledgment.
B. I did not make any outward acknowledgment.

Later in the response sheet, you provide further information about your fly experience: "How real was it to you? How vividly did you hear and feel it? Did you really believe at the time that it was there? Was there any doubt about its reality?"

As it turns out, so dutiful and respectful were the students toward me, their foreign visiting professor, that they deemed it impolite to circle choice B or not to embellish their experience with the imaginary fly. "After all," one of the students later admitted to me in confidence, "you've traveled a great distance to share your knowledge with us. That's the least we could do." In practice, however, this type of deferential response across the items of the Harvard scale inflates the measurement and skews the results.

I shared my observations with my local host. I also commented on the fact that everyone seemed very interested in the topic, yet not one student asked a question at the end of my talk. My host, a soft-spoken Singaporean woman, offered a bashful explanation. "When you lecture, you speak very fast, in a loud and authoritative voice. You are a large male and move your hands a lot... The students may be too intimidated to ask a question. They probably don't want to say the wrong thing to upset you. I think they may also construe your style as a bit aggressive. We are a small country, you know, and people around here usually keep their voice down..."

It's situations like these that show how the Harvard and Stanford scales are hardly optimal when it comes to measuring suggestibility across different societies and communities; cultural intelligence counts.[48] When it comes to hypnosis, for example, scholars assume the tests are translatable. This approach, however, operates under the specious assumption that the concept of hypnotizability is universal across cultures. But it's not that simple: whereas translations likely conserve the linguistic content, they may imply different meanings to people of different backgrounds and geopolitical subtexts. For example, Japanese culture shrouds hypnosis in unprecedented layers of taboo. In Japan, hypnosis refers to the macabre and supernatural;[49] moreover, in line with Japanese horror folklore, it signifies an immensely malevolent encounter.[50] In other words, both culture and brain shape the construct of hypnosis and suggestion, as well as its experiential and behavioral expressions. Subsequently, Western instruments are subpar tools to measure hypnotizability in non-Western populations—most communities in the world—even if they may provide necessary knowledge for the culturally sensitive reformulation of such tools.[51]

*   *   *

Suggestibility yields information that can greatly increase both diagnostic accuracy and treatment efficacy. If you are highly suggestible, you will likely benefit from certain kinds of therapeutic interventions where less suggestible individuals would not. Moreover, the availability of scientific measures of suggestibility has facilitated the proliferation of an impressive body of research and the general recognition that suggestion, including hypnotic suggestion, is a valid and vital area for promoting both clinical intervention and scientific inquiry.

Beyond embodying an important notion throughout the history of psychology and psychiatry and a central concept to the historical development of hypnosis,[52] suggestibility has been used to explain placebo response[53] and personality characteristics.[54] Are highly suggestible individuals better at responding to placebos? Can the suggestibility of individuals pave the road to recommending specific interventions over others for them? Whereas earlier efforts examining the relationship between suggestibility and placebo response found little overlap,[55] this view has been slowly changing as a function of our increased understanding of the science of placebos and the therapeutic encounter.[56]

Some professionals feed on and work with suggestibility as part of their day job. Using interrogative suggestions and suggestibility, for example, lawyers entrap suggestible witnesses in court, and using imaginative suggestion/suggestibility, clinical psychologists benefit their patients in therapy. They capitalize on a form of "situational suggestibility." In the context of their practical approach, if I can influence you (say, without your consent or knowledge) and implant an idea in your mind, then you are suggestible. In this sense, people sometimes construe being suggestible as a tendency for submission or unconscious compliance.[57] Stage hypnotists and magicians also know about this situational suggestibility. When interacting with members of the audience, good showmanship and stage presence depend on the ability to assess and tap suggestible volunteers quickly and effectively.

On the flip side, we can look at suggestion from the perspective of the influenced person, who assumes an imitative mental attitude while mistakenly thinking that everything is happening naturally. Even within a limited social interaction—a brief interrogation, for example—people can accept messages communicated during formal questioning; subsequently, these individuals may change their behavior in line with these suggestions.[58] In the context of memory, suggestibility often refers to a situation whereby people come to accept a piece of post-event information and incorporate it into their recollection.[59]

I am especially fond of situational suggestibility and have used it extensively in my magic and my science both within and outside of hypnosis. The main point in such situations is to leverage some unusual effect through a carefully crafted situation. Let me share with you some of my favorite libation stories.

Suggestible Wine

Not too long ago, having drawn some blood for a physical checkup, my friendly primary care physician took a long, hard look at the results and suggested I add a glass of red wine to my meals as a preventative measure to lower my bad cholesterol. "I know you don't smoke and hardly drink," he said with a wink, "but a bit of red wine is better than not drinking at all. It may even help you live longer."

Many people would probably embrace such a delightful medical recommendation. After all, if it makes a positive contribution to our health in addition to culinary and cultural sense, we may be witnessing a rare moment when doctors' orders align with the stars. But if you look into what rigorous metascience shows, you will find that even the moderate drinking of alcoholic beverages bears no health benefits.[60]

While wine, especially red wine containing polyphenol compounds, has garnered a reputation for having health benefits,[61] newer studies have found that even modest consumption of alcohol may contribute to health problems. In January 2023, the Canadian Centre on Substance Use and Addiction issued new guidelines warning that no amount of alcohol consumption is healthy and urging people to avoid it. Although American dietary guidelines are hardly as strict as the Canadian ones, numerous other health organizations have amended their recommendations to include the proviso that people should not drink alcohol for the express purpose of improving their health.[62]

I decided to discuss these developments over dinner. My dinner guests, all wine aficionados with medical training, were a bit surprised when I asked them not to bring a bottle of wine for our meal. I explained that because the supposed health benefits of drinking alcohol were not significant, I'd like to shut down the family bar and drink up the whole stash of mostly red wine that had accumulated over the years. What better way to do so than to have wine mavens around? As a closer, I mentioned that our family collection contained some wine gems, including a bottle of well-aged, top vintage Château Margaux.

During dinner, our conversation naturally veered toward discussing medical and scientific research concerning alcohol. At one point, someone mentioned that the price of wine influences how people experience it: expensive wines can seem to taste better than cheaper ones.[63] I took out a few bottles of red wine from the bar and challenged my friends to independently taste and rank them from most to least appealing so that we could compare them with their respective price tags. Dinner had turned into an informal experiment.

In the privacy of my kitchen, I poured the red wines into identical glasses labeled only with indifferent numbers: one, two, three, four, five…And my guests went into a spree of swooshing, sipping, gargling, raising glasses to the light, and making lip-smacking motions as they evaluated the assorted samples. This grading process included reliance on terms I had only passing familiarity with—turbidity, millésime, smaragd, and Weissherbst. The results were soon in, and some erudite dinner conversation followed. But not one of the high-minded wine connoisseurs was able to call my bluff: one of the wines was a white wine into which I slipped a bit of flavorless red food coloring to make it pass for a red…In fact, it got some great reviews, and a few even identified it as the Château Margaux![64]

Not only was this informal "experiment" hardly a novel idea, but I felt compelled to try it because of its cheek and innovation.[65] In the original study,[66] the researchers first gave twenty-seven male and twenty-seven female participants a glass of red and a glass of white wine and asked them to describe the flavor of each. Whereas white wine descriptors included terms such as "floral," "honey," "peach," and "lemon," red wine commentary elicited "raspberry," "cherry," "cedar," and "chicory." A week later, participants came back for another tasting round—again, a glass of red and a glass of white. However, unbeknownst to them, the two wines were actually the same white wine from the previous week, only the researchers laced one batch with some tasteless red food coloring to have it masquerade as a red wine. Subsequently, participants gave the white wine descriptors similar to how it had been described in the first tasting, while the white wine dyed red received commentary congruent with that of red wine.[67]

Evolution crafted the human brain to process gustatory and olfactory information about an order of magnitude slower than visual information—roughly four hundred milliseconds versus forty milliseconds. Evolutionary fitness likely prioritized spotting a predator over smelling danger. In the hierarchy of perceptual sensory input, vision takes precedence. But here, it misleads the participants in their ability to judge flavor. So strong is this visual illusion that even when participants intentionally smell the wine, deliberately assess its odor and flavor, and carefully verbalize their response, they do so in line with what they see because the color of the wine and the context of the dinner seem to frame the underlying sensory and cognitive process.

Strange top-down influences can also drive the *effects* of alcohol. One of my now-retired colleagues from McGill University once related to me an account from the time he was a graduate student. He lived in a university house for married students, which was adjacent to the football field on campus, and wanted to take advantage of his prime location. He figured he would be able to make a bit of money by stocking up on beer in his basement and then selling it during games to those who would order in advance, for a cheaper price than that of the stadium vendors. Accordingly, he sold beer tickets before the next game. He was about to stock up his basement when a letter from the dean advised him that the university did not permit any students to sell alcohol on campus. Thinking quickly on his feet, he bought a whole batch of nonalcoholic "near beer" and held his breath as he sold the "booze" during the game. Much to his amazement, not only did his happy customers not complain, but some ended up showing symptoms of severe intoxication, including loss of balance, vomiting, and slurred speech. My colleague survived the dean unscathed but promptly discontinued his entrepreneurial ambitions for psychological research into the effects of alcohol.

Moreover, this outcome mirrors a published study that observed the effects of expectation on social drinkers. Some participants got vodka tonics, while others got tonics with lime. Researchers truthfully told some participants what they had received, while they led others to believe that they had received the opposite drink. Amazingly, those who had received the tonic water but were under the impression that they received a string of the vodka tonics actually behaved as if they were intoxicated. This "Think-Drink" example illustrates that our beliefs concerning how we are supposed to behave after downing alcohol shape our actual behavior.[68]

We know that alcohol provides humans with both pleasure and pain. In America, the pain includes drunk driving and lost lives but also addiction, family strife, crime, violence, poor health, and squandered human potential; every American is paying for alcohol abuse.[69] The costs of alcohol-related harm—including expenses related to health care, law enforcement, and losses in workplace productivity—have dwarfed alcohol tax revenues.[70] Alcohol has killed more people compared to heroin, cocaine, meth, and prescription opioids even before we account for its contribution to fatal falls, deadly car crashes, and drownings. Excessive drinking costs America about $250 billion per year.[71]

We all want to drastically cut the number of deaths on American roads, spare many from fatal diseases, and reduce violence, sexually transmitted diseases, and unwanted pregnancies. When a society begins to appreciate that psychological processes have as much—sometimes more—to do with some drinking behaviors as do the physical effects of alcohol, we can begin to apply these insights to alcohol consumption and other addictive behaviors. Unfortunately, such approaches tend to elicit shrugs at best and resistance at worst from the clinical community. This dynamic happens, at least in part, because these psychosocial arguments seem to challenge a biocentric view of the scientific basis of medical practice. Still, the behavioral and social sciences are sciences too;[72] specifically, psychology is a science.[73]

* * *

The National Geographic producer spoke enthusiastically over the phone: "We'd like to feature the power of suggestion on our next big TV series, *Brain Games*. We know about your research and think you'd be the perfect brain expert on the program. Can we book you for our next shoot in Las Vegas?"

As a young performer, I would have quickly dropped everything to spend a weekend in Vegas. A few decades later, my academic career and growing family had shifted the balance toward a less itinerant lifestyle. In this case, however, I was happy to swap the frigid cold of my home in Montréal for the dry heat of Las Vegas. Plus, I like Nat Geo. As a university professor, I felt it was within my professional mandate to contribute to their creative educational content. So I accepted.

Arriving at the Bellagio, I reconnected with a magician friend who asked whether I was in town to play the role of a magician-scientist or that of a scientist-magician. I was a bit confused by his question. I told him I wasn't sure whether Nat Geo even knew I was a magician; they had tapped me as a neuroscientist. On my way up to the room, however, I got a disquieting call from the two producers. Before I could break out my short-sleeve tee and sandals, I learned that the segment taping would occur in a high-end ice bar.

Against the twenty-four-degree Fahrenheit landscape of arctic sculptures, ice-glass brain dioramas, and other over-the-top props emphasizing the subfreezing temperatures, the young producers wanted me to hypnotize random folks. "Can you make them believe they are burning hot?" they asked. This creative duo envisioned people stripping their clothes off until they were scantily clad. "In an ice bar, this visceral image would make for a *cooool* visual! Don't you think?"

I was disappointed. Not only had I exchanged the chilly Montréal winter for an auditorium-sized Vegas freezer, but the whole thing was also starting to feel more like sensational showbiz and less like the science education I thought I had signed up for. My magician friend had been right.

The producers had already selected their participant. "She is a tough cookie," they announced triumphantly, "if you can do her, you can do anyone." I was surprised by their choice of the verb "do" to mean hypnotize.

After a good coating of makeup and donning a heavy arctic-white fur coat, I was sure that the whole production was over the top. Just before the smoke machine kicked in and the cameras started rolling, I met Ali. Although she and I had never met before, I could intuitively sense that Ali was a bit anxious and tangibly excited about our on-camera session. "Have you ever been hypnotized, Ali?" I asked. Noting her silent head shaking from side to side, I began to take her through the motions of hypnosis. When she was ready, I made my first suggestion: "You will begin to feel very warm." As I continued my incantations, she began to strip away layers of clothing, one at a time.

"Cut!" the content director eventually thundered, having stretched the session to the point that any more exposed skin would get the Federal Communications Commission to flinch. Some folks on the set burst into spontaneous applause; a few encircled me and started to ask questions. One persistent theme resurfaced: "So, is this hypnosis stuff real? Do hypnotic suggestions actually work?"

I was trying to understand why I was answering these questions off camera. "Don't you want me to answer these questions for your viewers?" I asked the producers. In my mind, that was the crux of the program. But the producers, whispering and huddling over a monitor, were too busy discussing some issue and left my question unanswered.

As per my increasing suspicion, it became painfully clear that this dynamic twosome had chosen Vegas for all the wrong reasons: they had conflated street and stage hypnosis—the purview of magicians and performers—with the medical and psychological hypnosis that practitioners and scientists use in clinical settings and behavioral research.

I tried to protest and elucidate the difference between "clinical hypnosis"—in other words, using hypnotic suggestions to help with medical and psychological conditions such as stress, anxiety, reduction of pain and bleeding, quitting smoking, improving eczema and warts, surgical discomfort, or the promotion of a general sense of healing—and the entertainment of "stage hypnosis," when the audience enjoys viewing the live, histrionic behavior of volunteers who step on stage. However, it was a lost battle; everyone (but me) viewed this experience as an engaging magic show, and I was the raucous magician.

We had a good chat when the producers finally finished their consultation and got back to me. The conversation took an interesting turn: "What you did with Ali was very impressive," one of them said while the other kept silent, scratching his head. Then scratchy spoke, "*We* know that Ali was not your shill, but how would our *viewers* know whether she was playing things up for the camera or if your suggestions moved her?" I was about to respond when he scratched a bit more: "Would another person have responded the same way, or was Ali special?"

I was pleasantly surprised by this line of critical questions. We were getting closer to where I wanted things to go. "You'd like to know what a brain scanner would show if we could peek into her mind or glimpse the workings of her inner consciousness throughout this demo?"

The dual directorship vigorously nodded their heads.

The professor in me wanted to retort with the academic "It's complicated," but my internal magician-scientist had a better reply, and so did the scientist-magician. I decided to share the secret to Ali's hypnotized behavior with the directors. But before I could explain how the gentle binding of suggestion with imagination works and why it can change our mind, I had to make sure they understood the history of how it came to be. They needed to know about the evolution of suggestion.

CHAPTER 2

The Evolution of How and Why
Humans Are Suggestible

When my eldest child, Glenn, was in middle school, he was already curious about some of the research going on in my lab. He started hanging out at the Brain Institute, a research hub I founded at Chapman University in Southern California to advance research questions with experimental paradigms and scientific equipment; soon thereafter, he started to take university courses after school. In the summer of his eighth-grade year, he traveled with me to work every day to conduct his own teenage pursuits interacting with my university colleagues, college students, and researchers.

One afternoon, Glenn showed up in my office with a pale visage, apparently distressed. "What's up?" I probed, expecting something was amiss.

"I was making lunch in the kitchen," Glenn mumbled with some red schmutz visible on his upper lip, "and planned to have a can of baked cannellini beans in tomato sauce. But when I tried to open the can, the ring broke off, and the lid got jammed.

"I couldn't open it," he continued. "Whatever I tried, the lid wouldn't budge. I didn't want to bother you, so I got a big knife out of the kitchen drawer and tried to pry open the can like a lever..."

I stopped what I was doing and turned my full attention to Glenn; whatever he was trying to tell me seemed important and involved a large knife. I quickly scanned his hands, but they seemed intact and without cuts. No signs of injury except a few more red stains to match his upper lip. "What happened then?" I inquired irritably, trying to quickly get to the bottom of his distress.

"Well," Glenn answered unhurriedly, "after a few minutes, I was able to get the lid off, but it was a struggle, and some sauce spilled out..." Impatient with his slow delivery, I hustled to the kitchen to inspect the damage myself. There were clear signs of battle on the counter and the conference table outside, along with a kitchen towel drenched in tomato sauce.

"Better clean up," I said, a bit unsure why my young teenager would suddenly care about the mess he was leaving behind.

"It's more complicated than that," Glenn retorted. "When I ate the beans, after the first or second bite, I felt something hard when I was chewing, but it was too late, and I swallowed it..."

"Something hard?" I asked with surprise, "You mean a hard bean? A filling? A tooth?"

"Actually, I think it may have been a metal piece." Glenn looked at me helplessly.

"Why do you think it was metal?" I probed in disbelief.

"Because after I swallowed it, I walked back to the kitchen and looked at the knife... The tip is missing, you see?"

It was a large chef's knife I had never seen before. I had recently bought the lab a whole bunch of cutlery from an estate sale—an entire box full of assorted kitchenware, including a few large knives for everyone to use around the break room and kitchen. Sure enough, the knife before me was without a tip; truncated, the blade came to a dull end with a small, pointy triangular piece missing.

"Are you saying the tip broke off while you were prying the lid, fell into the beans, and then you ate it?" I asked with incipient alarm.

"I think so," Glenn replied weakly. "I am not 100 percent sure, but I think so…"

I took a long, hard look at my son and at the knife, disbelief racing through my head. On the one hand, if he *had* swallowed the sharp tip, he ran the risk of perforation to his internal organs. I'd need to rush him to the pediatric emergency room without any food or drink, get some x-rays, and possibly scope out the shred. On the other hand, this used knife may have been dull from the get-go. Glenn wasn't showing any symptoms—no fever, no swelling—other than his confused anxiety about this situation…

"How are you feeling, buddy?" I asked, putting on my clinician's hat and ascertaining his breathing and vital signs.

"Fine," Glenn replied, but his voice wavered unconvincingly.

"Any pain anywhere? Difficulty breathing? Did you choke? Gag? Faint?" I asked in rapid fire while examining the remaining beans and sauce to look for any clues or remaining shards.

"I sort of feel a sharp point in my throat about here," Glenn said while pointing to his Adam's apple. I gently touched the familiar projection in front of his neck but knew that without x-ray vision, I couldn't probe behind the soft cartilage of his larynx.

"Does it hurt?" I asked gently.

"Only when I swallow…" he replied with a sense of terror in his voice.

I made a quick phone call to my wife and shared with her my calm analysis of the situation. But my clinical evaluation was incongruent with the sensibilities of a mildly frantic mother: "Get him to the hospital," she demanded, "right now!" I hung up the phone only to discover that Glenn was clutching his throat and struggling to swallow. Without another word, I started the car, and we hightailed it to the emergency room.

I presented Glenn and his throbbing throat to the triage team. However, they were moved neither by the story nor by the large tipless knife I pulled out of my bag as evidence. The imaging technician came by to take some x-rays of the neck and chest.

"How's it looking?" Glenn blurted out from under a pair of big, concerned eyes.

"I'm just the x-ray tech," the guy replied, keeping a poker face. "I'll show your images to the doctor, and they will get back to you, buddy. Hang in there."

"I was hoping he'd be able to tell me..." Glenn whispered to me as soon as the tech left. Looking at my distressed son, I could see his symptoms were becoming more serious: difficulty swallowing, heavy breathing, and persistent, sharp pain in his throat.

When the ER physician finally entered the room, she first examined Glenn without saying anything for a few seconds. "Where does it hurt?" she asked in a soft voice while palpating his neck gently.

"I have this throb right here," Glenn pointed to the laryngeal prominence of his thyroid cartilage and looked at me for help. Again, I pulled out the mighty chef's knife and presented it to the ER doctor while concurrently reciting story highlights.

She looked at me briefly, peeked at the truncated knife, and then politely interrupted my narrative with a weary smile: "You're a doctor, so we can cut to the chase," she said dryly while handing me a heap of paperwork. "The x-rays were negative."

"What does it mean?" Glenn asked me after the door closed behind her.

"It means you have no foreign object inside your body," I answered.

"Is it 100 percent certain?" Glenn asked with distrust.

"Yes"—I nodded with a fusion of relief and exasperation—"if you had any metal bit in your body, it would light up the x-rays like a Christmas tree!"

Glenn looked a bit confused but then broke into a heartening smile. "You know what, my throat—the pain's gone. I can swallow smoothly!" And just like that, Glenn was back to his calm, suggestible self.

As Glenn and I headed back home on a much more leisurely drive, it seemed reasonable to mull over the potential evolutionary benefits of being suggestible. Specifically, two pressing questions come to mind: Is there an evolutionary advantage to suggestibility? If so, what could it be?

The Evolutionary Advantages of Suggestibility

Let's start with the overarching answer: yes. Suggestions occur ubiquitously in ordinary life and have evolved from an ancient means of interaction that did not always require spoken language. A process of communication and learning resulting in the transfer, and possible acceptance, of proposed information is important for members of any social group.

In addition, from a clinical standpoint, suggestion and suggestibility play a role in the evolutionary perspective associated with curing and healing, including through placebos. For example, the trust of a sick individual in a traditional healer—doctor, shaman, or wise sage—may improve survival and provide an evolutionary advantage. Moreover, suggestions play a role in the transmission of ideas, meanings, beliefs, symbols, and behaviors from one generation to the next through the construct of culture. As outlined below, suggestion and culture work closely together to bring about an evolutionary dynamic that works more quickly than genes. Even today, no human—regardless of how learned and transcultural—is completely immune to suggestibility.

I started thinking about the evolution of suggestion around 2000, as I was transitioning from a bachelor at Cornell Medical Center to a married man at Columbia University College of Physicians and Surgeons in the City of New York. However, these insights came from a surprising source—the American Museum of Natural History (AMNH).

Living a short walking distance from the AMNH on the Upper West Side of Manhattan, I decided that rather than sign up for a gym membership, I'd get my physical and mental exercise by regularly walking the halls of this great institution. It became one of my favorite pastimes, and I cannot even begin to calculate how many miles and flights of stairs I have since traversed within this iconic establishment. With increased familiarity, my repeat visits taught me where to go and when to avoid the mass crowds of school kids and tourists. The AMNH became, and has prominently remained, my favorite museum ever.

I studied the collections there—old and new—systematically. Over the years, I read explanations, viewed artifacts, followed docents around, listened to lectures, played with interactive demos, watched educational videos, attended special exhibitions and workshops, and marveled at the vast compilations of knowledge the AMNH offers, including the treasure trove of materials in its remarkable library.

When Glenn was born, I used to take him with me on my AMNH excursions—first in a baby carrier, strapped around my shoulders, and later in a stroller. He, too, became very familiar with the museum and the wonderful peek it offers into science and evolution. Baby Glenn quickly developed a clear preference for visiting the Hall of Primates on the third floor. No visit to the museum would be complete without saying hello to our evolutionary "cousins"—chimpanzees and bonobos. Glenn loved to stare at the range of primates and their relatives—from rodent-like tree shrews to mighty gorillas; from spider monkeys with their long, grasping tails to tailless apes with hands specialized for swinging from trees. He had an inordinate fondness for this display.

During our visits to the Hall of Primates, I'd read out the signs to him; together, we explored the relationship of humans to other primates.

He was too young for the technical details—say, that chimps and bonobos are our closest living relatives, with whom we share nearly 99 percent of our DNA—so I just paraphrased the explanations in age-appropriate vernacular and loosely referred to "chips and bobos"—to use his pronunciation—as our "cousins." This glib approach got me in trouble when Glenn met my sister's kids—his real cousins—but by that time, the AMNH had opened the dazzling Anne and Bernard Spitzer Hall of Human Origins on the main floor. Subsequently, Glenn and I shifted our attention to the long arc of human history, from early ancestors who lived more than six million years ago all the way to our own times.

Before long, Glenn could quickly parrot that our species, Homo sapiens, evolved in Africa about two hundred thousand years ago. He probably had a shakier handle on the idea that modern humans hail from this African population of Homo sapiens, which continued to develop in Africa but then spread out—for example, to Eurasia—some fifty to one hundred thousand years ago. I tried to explain to him that while humans were busy colonizing the globe, we interacted with earlier human populations—Neanderthals, Denisovans, and perhaps other archaic species that we met along the way.

Although I did my best, it was difficult for me to explain to young Glenn that this "interaction" included sexual exchanges and extensive interbreeding between our species and those local populations. With time, this procreative dynamic diluted these earlier human populations until it completely replaced them; now we—modern humans—are here, and they are extinct. Although we are the only surviving human species, we carry some of their genetic contributions within us. To be sure, 2–4 percent of our genome today still shows these traces.

"Is part of me very old?" Glenn asked when I explained that DNA studies show Neanderthal and Denisovan genes in all living non-African human populations.

"Sure!" I answered with my eyes wide and eyebrows outstretched. "That's a great way to think about it!"

In this regard, suggestion and suggestibility are evolved traits, not unlike bipedalism or three-color vision. They have emerged over time, and we need to think about them in that context. We have been carefully crafted to be suggestible—some of us more, some less—likely because suggestibility serves an evolutionary purpose.

* * *

A few months later, during a typical AMNH run to the "cousins," Glenn suddenly raised his empty sippy cup, turned his head away from chips and bobos, directed his inquiring brown eyes at me, and mumbled in a high-pitched voice through the noisy room: "What's special about us?"

I took his precocious question to mean a variation on "What is unique to humans?"—compared to other species.

As I organized my racing thoughts to tell Glenn about our special human features, I struggled with how to broach it with his tender mind. Should I describe how we have large brains relative to our bodies and our increased ability to leverage speech and sustain movement on two legs? Should I point out our short, blunt canine teeth, their eruption before the premolars, and our projecting chin? Or would it make more sense to talk about our sparse body hair, permanent enlargement of breasts and buttocks in mature females, short, straight finger and toe bones, prolonged thumb for precision grip, or our lengthy period of development after weaning, protracted adolescence, and average life span that is longer than that of other apes? Somehow, none of these points seemed easy or appropriate to discuss with my stroller-bound toddler.

Instead, I wondered whether I should skip straight to our special ritual of burial of the dead, worldwide presence, skill with crafting and using complex tools, composition and performance of music, written and spoken language, art and law, rhetoric and logic, and symbolism. Amid my contemplation, his little voice interrupted my stream of consciousness.

"What's 'special A+'?" he repeated.

He'd probably picked up "special A+" from the droves of competitive elementary and middle schoolers who boisterously shared the museum with us attempting to ace their AMNH worksheets. Not quite "What's special about us?"

But I had already thought of another something that is uniquely human: the transmission of cultural ideas, meanings, social interactions, and behaviors from one generation to the next—something that pivots around the construct of suggestion. For example, for the Aboriginal people in Australia, songs serve as maps of the land and describe a myriad of helpful information, including geography, ecology, hunting behavior, community norms, trade, history, and rituals. These songlines are transmitted orally from generation to generation through songs, stories, and dances. Within a broad context of cultural heritage, they serve as musical suggestions that members of the group chant together, discuss, and reflect upon.

Suggestion Facilitates the Cultural Evolution of Modern Humans

Any discussion of evolution logically draws on the biological sciences. Anatomically, we, as modern humans, continue to change even today. For example, consider the median artery—the main vessel that supplies blood to the forearm and hand. After it forms inside the mother's womb, the median artery usually disappears in lieu of two arteries that develop to replace it—the radial and ulnar arteries. Because this developmental process occurs during pregnancy, most adults don't have a median artery. But the presence of the median artery in adults has been increasing, from approximately 10 percent in people born in the mid-1880s to approximately 30 percent by the end of the twentieth century.[1] Today, the median artery is present in 35 percent of people, and researchers predict that all newborns will show this feature in a few decades. This microevolutionary change represents but one example of incremental formations in the internal anatomy of our human body. Moreover, when its prevalence surpasses 50 percent, the median artery will no longer be a "variant"; instead, it will become a "normal" human structure.[2]

We can speculate about some of the implications of having a median artery. On the one hand, a median artery may compress the median nerve to increase susceptibility to carpal tunnel syndrome. On the other hand, having three arteries is advantageous not just because it increases overall blood supply to, and the associated control of, our forearms and hands, but also because it can replace other parts of the human body in some surgical interventions.

Many people incorrectly think that modern humans have formed and that we have now reached a point where we no longer evolve. But rather than slow down or altogether stop, evolution accelerates and depends on our living environments, which are changing more rapidly today than in any past period. Today, one of the main vectors driving evolutionary change in modern humans comes from an overlapping but separate strand of evolution—one that scantily relies on biology alone, instead also drawing on social factors: culture. Suggestion and suggestibility play a key role in cultural evolution.

Whereas many people appreciate that genes, biology, and physical features play a substantial role in evolution, fewer understand that so do culture, society, and the environment. If you think of the theory of evolution as an attempt to explain the place and trajectory of species in the world over time, then the direction in which humanity is going is less a function of changes in genetics than a function of variations in culture. Even if we ignore for the moment the differences between various types of suggestion and suggestibility, it should become clear that such constructs permit us to respond quickly to changing cultural, societal, and technological demands. In this way, suggestion and suggestibility allow us to adjust speedily to social trends and cultural changes.

Although evolution classically emphasizes "natural selection" and "sexual selection," it also anticipates the third primary instinctual process— social relationships. When we observe ants, bees, and other social species that, like us, thrive on group interactions, we recognize the importance of a communal instinct, which seems as important as survival and sexuality. Humans are social organisms; as such, we are only as strong as our communities. In this regard, suggestion serves as a major lubricant in our ability to effectively interrelate, cooperate, and get along with others.

Cultural Suggestions

Culture has a considerable influence on our suitability for the environment,[3] and cultural impact on evolution is faster and more potent today than ever before. Culture has become embedded in a variety of qualities within the "skeletal backbone" or "life history" of our human core.[4] This form of cultural evolution shares parallels with, as well as differences from, genetic evolution.

For example, consider human height, which serves as an index of the health and well-being of an individual and a population. Height variation results from a complex interaction of genetic, environmental, socioeconomic, and cultural influences. Because a calorie used for growth cannot be used for fighting stress, data show that adult height, when used as a measure of child growth, is an indicator of a stressful environment in context with genetic background and spatial factors. Specifically, stress and social factors exert a greater effect on the height of men than on that of women: stress affects the behavioral, endocrine, and molecular responses of brain systems in the hypothalamus—a key node between the endocrine and nervous systems—and this effect presents itself in a sexually dimorphic way, with males being more vulnerable.

The parameters for short stature, beginning with the most impactful, include income inequality, air pollution, gross domestic product (GDP) per capita, corruption perception index, homicide rate, life expectancy, and unemployment. In other words, in environments steeped with more stress—say, countries with a high level of violence—people tend to end up with smaller physical stature; moreover, parameters such as corruption and economic inequality correlate with the height of the population.[5] Now you know why Scandinavians are so tall: they don't need an explicit suggestion to be so; they have the genetic background of tall stature from their Viking ancestors alongside a low corruption rate and high GDP!

So, cultural and genetic evolution cooperate and interrelate to result in gene-culture coevolution, in addition to purely cultural evolution. This dynamic holds societal implications for the ways we interact with one another, as well as with our environment.[6]

* * *

We often mistakenly believe that we evolve and change as a way to adapt to our environment, but in fact, we also change our environment to suit our own needs. Beavers, for example, can build dams and lodges to form lakes that stretch for miles and change the ecosystem, sometimes for decades.[7] Accordingly, scientists are beginning to talk about a new kind of inheritance system parallel to our biological system of passing genetic material: a system that inherits the environment.[8] The gophers in my backyard use a system of underground burrows that they inherited from their predecessors. Despite or perhaps because of my protests, they further developed it, deepened their network of lateral tunnels, and continued to thrive in my backyard. Moreover, they will gift this precious infrastructure to the next generation of Raz backyard gophers.

As humans, we, too, inherited the environment from our ancestors, changed it a bit (to say the least), and now pass it on to our offspring, who will continue to cultivate, preserve, and pass it on to future generations, if, of course, the extensive changes we've made haven't already destroyed the environment we live in for the next generations…but that's another story.

The behavioral phenotype of activities such as the nests termites build or the burrows gophers dig follows an essential concept in evolution—niche construction. A particular case of constructing a niche is building a cultural niche.[9] For beavers, building dams is instinctive.[10] For example, if beavers in a cage hear the sound of running water, they will get intensely busy building a dam, even if the cage is completely dry.[11] Humans are also intensely busy with constructing niches; however, whereas beavers do it innately, for us, knowledge is primarily *cultural*, not inborn, and often driven by suggestions (recall the songlines of the Aboriginal people of Australia).

Unlike beavers, we build as a result of our cultural heritage, not our genes. Consider prehistoric humans—a small portion dwelt in caves, but most didn't. Caves are difficult to find across most of our planet; they are dark, cold, damp, and often already occupied by unfriendly tenants, from bears to lions to hyenas. And yet, in the case of early humans and Neanderthals, who were around until several tens of thousands of years ago, they didn't erect fences or build walls, not even at the crude level of piling up a few stones to protect themselves from the harsh winds. Although this type of architecture would have been helpful, it only appeared much later. Why? Because without exposure to a cultural suggestion, it's difficult to realize it. If you've never come across—and nobody has ever suggested to you the idea of—railings, barriers, and shelters, you're unlikely to build.

The emergence of the "Let's Build" concept forms an evolutionary inflection point. Although cognition levels today and back then were broadly comparable, what young kids can effortlessly do today with Legos our ancestors couldn't possibly do even with their best efforts. Our predecessors didn't think of it because, from an evolutionary perspective, exposure to the physical environment isn't enough; suggestive exposure to the cultural environment is just as crucial. In this regard, suggestions are critical in the transference and learning of cultural information.

Throughout our evolution, different cultures and religions have long capitalized on human suggestibility. In addition, the effects of contemplative practice, prayer, food and fasting, alcohol, sleep deprivation, and psychedelics propose that rites and rituals may help the mind heal itself by enhancing suggestibility. Entrenched in suggestion, such practices—like modern-day cognitive behavioral therapy—may lead to better therapeutic outcomes by strengthening metacognitive intentions and group interactions with the aid of suggestibility as a mediating factor.

Specifically, scholars and clinicians have long been interested in these themes; however, the relatively short history of modern scientific and medical research has made it difficult to study the effects of suggestion on humans. It is curious to notice how this research trajectory has emerged as a derivative of the way belief influences clinical outcomes, as well as deception and trickery.[12]

CHAPTER 3

Do Honest Scientists Ever Deceive in Research?

D eceit is in the eye of the beholder. People can feel deceived when they actually weren't, or they can feel that they weren't deceived when they actually were. Things become even thornier in the world of research. The ethics of using deception in research force us to think long and hard not just about moral principles, but also about what constitutes reasonable deception and what counts as legitimate research.

Dark Dreams

My personal introduction to ethical problems in medical experiments came early. Although I was only five or six years old, I remember this experience clearly. It would happen nearly every time I went for a sleepover at my grandparents' place. I would be suddenly awakened from sleep during the first half of the night by screaming coming from my grandparents' bedroom. My grandmother would then tiptoe into my room to check on me, calm me down, and tell me that everything was fine.

"But why is this happening?" I used to ask her every time it occurred.

"He just had a bad dream," she'd invariably say, "go back to sleep. I will see you tomorrow, bright and early, for a great breakfast."

It felt like a dark secret. I already knew that she wouldn't address any of my questions at breakfast. She never would; neither would my grandfather. To this day, etched in my memory is the image from when I sneaked into their bedroom and caught a glimpse of him: sitting up in bed, trapped in a fog—neither fully asleep nor fully awake—breathing heavily and whimpering, and difficult to comfort despite the care and reassurance that my grandmother would shower him with. His dramatic nocturnal awakenings would both startle and perplex me.

It took me years to figure out that these were not nightmares, but night terrors, and that my grandfather—a physician who survived World War II and thereafter immigrated to Israel—was suffering from trauma. He was experiencing the Holocaust vicariously through his patients in Tel Aviv; he was providing medical care for survivors, including individuals who underwent "medical experiments" by Josef Mengele.

An anthropologist and physician, Mengele was responsible for hideous "research," particularly targeting twins, at the Nazi concentration camp Auschwitz. At that time, twin research was the preferred method to tease apart how nature and nurture—human heredity and the environment—influence human development. Most of his victims were children, and he would transfuse blood from one twin to the other, amputate limbs and sew them onto the other twin, stitch two twins together to form Siamese constructions, contaminate one twin with infectious diseases using the other twin as a control, and many other such offensive "experiments." The twins would typically die from his inhuman procedures. If they didn't, he would often kill them anyway to afford himself a "proper" autopsy. Moreover, if one twin died from a disease, he would often "sacrifice" the other to observe the differences between the sick and the healthy. The few who managed to survive this horror and made their way to Israel spoke to my grandfather, and they ended up in his dreams.

Deceit, Consent, and Obedience

A Supreme Court decision from 1914 established the right of medical patients to not receive treatment without their explicit consent.[1] Later, the Nuremberg trials—which found Nazi medical officers guilty of atrocious and sadistic experiments on concentration camp prisoners—questioned the benevolence of some physicians.[2] And yet, even after the Nuremberg trials, many biomedical researchers continued to take the liberty of administering treatments by way of various ruses. Accordingly, many people have developed a general knee-jerk reaction to anything that reeks of either deceit or lack of consent.

The Nuremberg tribunal concluded that "the voluntary consent of the human subject is absolutely essential.... [Participants] should have sufficient knowledge and comprehension... of the subject matter... to make an understanding and enlightened decision." Moreover, since the 1940s, this sentiment has matured into the doctrine of "informed consent"—when a patient grants permission, in sound mind and intact cognition, after understanding well the possible risks and benefits of a proposed treatment.

Post-Nuremberg, however, many researchers still felt unconstrained by this legal language and continued to take shortcuts. Over two thousand years of history, some researchers have engaged in medical research that is morally repugnant[3]; in the US, for example, the infamous Tuskegee study whereby doctors intentionally let Black men with syphilis go untreated[4] still haunts the Black community and the world of medical ethics.[5] Subsequently, during the 1960s, federal regulators gradually adopted various rules, beyond the requirement of informed consent, to govern the research of biomedical scientists.[6]

At the same time, social research came under attack with loud objections flagging methodology involving deception as incompatible with informed consent. Regulators found such practice particularly troublesome when research protocols caused participants to experience anxiety, embarrassment, and other forms of mental stress. The experiment that drew the sharpest fire of these critics was one by Stanley Milgram in 1963. I got to visit his lab, in the basement of Linsly-Chittenden Hall, when I lived in the Hall of Graduate Students at Yale University some decades after he had passed on. I remember standing in his former research space and thinking how Milgram—testing more than a thousand men ages twenty to fifty who responded to an ad in the *New Haven Register*—was the first to highlight the power of social situations to dominate regular people.

Milgram—hailing from a family of Holocaust survivors—asked himself whether a Holocaust-like type of genocide could happen again. The trial of Otto Adolf Eichmann, one of the major choreographers of the Holocaust, further fueled his interest in the limits of human obedience and deference to authority. Like many before him, Eichmann seemed to clear his personal conscience by claiming that he was "just following orders." Only decades later, in *The Devil's Confession: The Lost Eichmann Tapes* (2022), did audio recordings surface to further demonstrate how profoundly antisemitic he was and how intentional and merciless his inhumane actions were.

As an accomplished researcher at Yale and under the guise of research into the learning process, Milgram told his naïve participants to administer progressively stronger electrical shocks to an unseen student in the next room—a confederate whose disembodied voice came over an intercom system—whenever he made a "mistake" on a learning task. In return, they received five dollars for their one-hour participation. As the shocks grew stronger, rising to a supposedly dangerous level, the confederate—his prerecorded voice played back by tape—would moan, cry out in protest, and even shriek in pain, begging for the experimenter to stop.

However, when participants wanted to break off, Milgram, dressed in a white coat, would insist that the experiment required them to go on. "Several participants," he reported later, "were observed to sweat, tremble, stutter, bite their lips, groan, and dig their fingernails into their flesh"—yet most of them obeyed his orders. Milgram concluded that when someone else is in authority, perfectly normal people are capable of obediently acting brutally toward others, a finding that might cast light on the dark behavior of many of the ordinary citizens of Nazi Germany.[7] Milgram won the 1964 award for social-psychological research from the American Association for the Advancement of Science.

Citing Milgram-like experiments, opponents of deceptive research argued that psychological science should impose tighter controls on researchers. But the powers that be continued to tolerate deceptive methods in social research—with some exceptions—in order to not interfere with the freedom to pursue knowledge.[8] The dilemma stemmed from the coexistence of two incompatible American ideals: a belief in the value of free scientific inquiry and a belief in the dignity of people and their right to privacy. Researchers could explore certain areas of psychology only by robbing participants of their right to informed consent; society could protect that right only by depriving researchers of the chance to cast light on important issues through deceptive experimental methods.

Following Milgram, federal regulations tightened again, and many disheartened researchers had to give up on exploring problems they considered important. At the same time, some investigators persisted in their quest to seek knowledge within a labyrinth of restrictions. One such example was the Stanford Prison Experiment (SPE) led by Philip Zimbardo in 1971. As part of the fifty-year anniversary of the SPE, I had occasion to discuss with Zimbardo how his efforts pushed the ethical limits of research.

After receiving ethics approval from the newly formed human subjects committee at Stanford, Zimbardo and his research team recruited participants—young men who responded to a newspaper ad—in return for fifteen dollars per day. Flipping a coin, the researchers randomly assigned participants into one of two groups: either guards or prisoners. The participants then enacted their respective roles, using uniforms and accessories, in a prolonged simulation of jail life in a for-show prison constructed in the basement of Jordan Hall at Stanford University.[9]

Within hours, most of the guards had become sadistic and abusive; moreover, because some participants started exhibiting dangerous behaviors, the researchers terminated the experiment within six days rather than the full two weeks as originally planned. Zimbardo and his colleagues concluded that the vile actions of the prisoners and guards stemmed from the social roles they had assumed.[10]

While Milgram deceived and Zimbardo didn't, the point of these studies is that situations—while they may start off as artificial and contrived—can assume a reality in the minds of participants. Whereas Milgram illustrated the power of social psychology in the context of obedience using deception, in the Stanford Prison Experiment, everybody knew it wasn't real: each participant was acting, dressed up, in a role, yet that role enactment became their reality.[11] Curiously, although Zimbardo and Milgram never overlapped at Yale, they did overlap for two years (1949–1950) at James Monroe High School in the Bronx. But Zimbardo, who had left Yale just before Milgram started there, was the one who actually remodeled the basement space where Milgram conducted his landmark experiment. Whether in the dungeons of Linsly-Chittenden at Yale or Jordan Hall at Stanford, these "basement studies" by Milgram and Zimbardo, as well as experiments conducted by others,[12] provide profound demonstrations of how social situations shape human behavior quickly and dramatically.

While the SPE made a valuable contribution toward prison reform, both academics and nonacademics widely criticized the researchers for having inflicted suffering on the participants in the name of science.[13] Thereafter, again, the federal government tightened the regulations governing biomedical research with human subjects and toughened the policies even further in 1974. Among other things, informed consent has become a stricter document containing formidable legal language. Moreover, the new regulations covered behavioral research as well. Even minor deceptive methods in social-psychological experiments were now unacceptable unless the Institutional Review Board (IRB)—an internal committee within research institutions designed to uphold the rights of human participants—explicitly approved. Indeed, researchers need to disclose all aspects of their methodology to the IRB prior to running any experiment.

After Milgram and Zimbardo, social psychologists who wanted to walk in their footsteps had to write detailed explanations of their intended research, defend their proposals before suspicious IRBs, and "dilute the impact" of their experimental situations (as in play them down); even so, their proposals often didn't meet with approval. In surveys asking social scientists whether the IRB had impeded their research, more than half said it had.[14] Accordingly, in the early 1980s, IRBs began to relax a bit and accept experiments deemed as "minimum risk," if the research "could not practicably be carried out" otherwise.[15] For example, one of my mentors, who was doing research on hypnosis at the University of Connecticut around 1980, had to meet with members of the IRB to teach them why using hypnosis in a research study was ethical. After he gave them an hour-long talk about what hypnosis is and isn't, they easily approved his studies.

However, even in their slightly loosened form, the regulations have made a major change in social-psychological research. Deception is still in use, but mostly only in limited form and allowed to involve only minor—some would say trivial—experimental stress and discomfort. Subsequently, investigators had to narrow the scope of their research. "We don't even consider experiments that would run into resistance," one of my Cornell mentors said to me. "Whole lines of research have been nipped in the bud." Obedience experiments of the ilk Milgram conducted are now beyond the pale, and so are research designs that would even briefly frighten or embarrass participants.

Tricky Research

Opponents of research involving deception (hereafter "Opponents") typically claim that deceptive research is unnecessary and that alternative techniques could serve instead. For example, they have suggested the use of role-playing in which participants imagine how they would behave in a given experiment if they were naïve to its real purposes. In many cases, they give themselves better marks than truly naïve participants would achieve. But we know that what people say or do in role-playing experiments is qualitatively different from what they really do. Much the same, disappointingly, is true of various other suggested alternatives.

Accordingly, both federal regulations and professional organizations reluctantly continue to acknowledge the necessity of deceptive research and to use the cost-benefit ratio as the operative ethical standard by which to judge it. Today, researchers must meet the following three criteria for any research involving deception: 1) that the question to be answered is sufficiently important; 2) that deception is necessary to answer the question; and 3) that the subjects are debriefed about the deception in a timely manner. Displaying American pragmatism at its best, this approach judges the morality of any case of deceptive research by its results.

While philosophical concerns about the use of deception in research are well-intentioned, researchers in the trenches find limited negative (psychological) effects on participants. Moreover, undesirable or lingering outcomes from interpersonal deception on participant mood and attitudes toward research typically resolve after a debriefing procedure—an explanation provided after the experiment.[16] When coupled with judicious experimenter training and experimental procedures, not to mention IRB oversight, these results propose that, although the use of deception is hardly risk-free, its application (when necessary) poses limited potential harm to participants.

Opponents claim that costs greatly outweigh benefits. Their evidence, however, is largely anecdotal, drawing on participants who, despite debriefing, have remained disillusioned or disenchanted with their experience. Sometimes, participants who go through a deceptive procedure may confront unflattering epiphanies; for example, they may discover that they are stingy, selfish, or simply that their political opinions are tenuous.[17] Opponents call this outcome "inflicted insight" and argue that foisting unsought self-knowledge on anyone—certainly research participants—is morally indefensible.

On the other hand, most social psychologists maintain that deceptive research entails little, if any, harm. This view builds on surveying many participants, who report that, upon looking back on their experiences, the scientific knowledge gained justified the deception imposed on them. In the case of Milgram, 84 percent of participants said afterward that they were glad they had taken part in it; only 1.3 percent regretted the experience, and the rest were neutral.[18]

But hardcore Opponents refer to this argument as a "utilitarian ethic." They reject deceptive research on the grounds that concern for human rights is morally superior to the freedom to seek knowledge and should take precedence over it. Deception comes in many flavors; one size hardly fits all. But even if deception does no harm, some Opponent scholars maintain that it serves individuals wrong in that it deprives them of their free choice. According to this view, cost-benefit calculations reduce participants to convenient tools rather than treating them as moral agents with free will. Moreover, Opponents construe deception in the clinic as compromising the relationship between the practitioner and the patient.

But deception can, in some circumstances, be both ethically permissible and achievable without serious harm to the doctor-patient relationship.[19] For example, if a patient has a treatable ear infection that causes untreatable nausea, a doctor could give them antibiotics for the infection and a placebo for the nausea.

I shared my ethical musings with my ear, nose, and throat (ENT) doctor. "How about that?" I asked him when we met socially one day. "Would you consider such a strategy?"

"Sounds intriguing," the ENT replied with no special interest in the topic, "but I doubt I'll ever do this..."

I wasn't sure whether this answer reflected ENT antagonism or a lack of personal rapport, so I decided to take these queries about deception to a friendlier physician. I called up Rose, an eye doctor whom I trust.

A modern version of Wonder Woman, Rose is a highly scheduled professional who speed writes faster than I can talk and regularly makes large quantities of delicious carrot soup for her many house guests (and patients)—all while working full-time as a busy ophthalmologist. When we speak on the phone, in the spirit of efficiency, she usually juggles a few more tasks. I asked Rose whether she'd consider treating patients with untreatable viral conjunctivitis—"pink eye," an inflammation of the lining of the eye—by telling them that they would receive nothing for the virus, but a placebo for their itchy, uncomfortable eyes, an option that has since been labeled "open-label placebo" (OLP).[20]

My question stumped her. "Why would I want to do that?" she asked as the paper-shuffling clatter of sorting through her mail came to a silent halt. It was a rare moment when she focused her entire attention on me.

"Because it's a much better option than lying to your patients, as in the story of Mr. Wright," I replied enigmatically, and quickly went on to relate his tale.

A report from the late 1950s describes a Mr. Wright who suffered from lymphosarcoma, a class of cancer.[21] He had tried multiple treatments, none of which had been able to cure him successfully. The prognosis wasn't good because the cancer had progressed to the point that his body was filled with orange-sized tumors. But due to his persistence to live, Mr. Wright researched a promising new experimental drug—Krebiozen. For some reason, he believed this drug would save him.

The treating physician, who couldn't see why Krebiozen would do anything, decided to indulge Mr. Wright and gave him

"Krebiozen"—actually a placebo—on a Friday; the following Monday morning, Mr. Wright's tumors were half their original size. Ten days following his first dose of what he thought was Krebiozen, Mr. Wright left the hospital, seemingly cancer-free.

"Wow," Rose interjected. I continued.

But then the scientific literature began reporting that Krebiozen appeared to be ineffective; Mr. Wright read the reports and fell into a deep depression, and his cancer returned. His doctor, a committed physician, decided to tell him that the initial dose of Krebiozen had deteriorated during shipping, making the dose less effective, but luckily, he had just received a new shipment of highly concentrated, ultra-pure Krebiozen. He then injected the clueless Mr. Wright with distilled water, and again the tumors began to disappear. Alas, two months later, the American Medical Association announced that Krebizoen was completely ineffective, and Mr. Wright lost all faith and died two days later.

"From all people I know, you should be one of the most critical of this anecdotal account from the fifties," Rose said as the backdrop sound of mail envelopes ripping open resumed.

"Of course I am," I said with grateful acknowledgment, "but would you consider deceptively using a placebo in the clinic as with Mr. Wright, or would you feel that the approach I described for giving placebos for viral conjunctivitis is more ethical?"

"I guess I'd feel comfortable telling patients that they'd receive nothing for the virus, and placebos for their itchy, uncomfortable eyes," Rose answered with guarded hesitation, "but I'm concerned that for physicians to give placebos to patients would be bad form under any circumstance, no?"

"Why would it be bad form, Rose, if you are telling them the truth?" I asked.

"That's really interesting. I never quite thought about things in this way," Rose replied in a pensive mode. "I'd certainly consider it favorably, if I had a judicious and ethical way of navigating this terrain."

Magical Experiments

Imagine an experiment in which I show you cards with human faces printed on them. I ask you to simply tell me whether you find the person in each photo to be more or less attractive.[22] Based on your responses, I form two face-down piles: one for more attractive and one for less attractive faces. So far, so what? Unknown to you, however, now I perform a magical move, which would escape the awareness of most people, and swap the two piles. I then reach for the "attractive" faces, which is actually the "unattractive" pile, and ask you to go through each card and explain why you found this person attractive.[23]

Strange things happen when this type of psychology experiment beefed up with magic takes place. When performed by a proficient sleight-of-hand artist in a way that interferes with the recall of even great visual memorists, participants will be convinced that these are their attractive selections and proceed to explain why they found these faces attractive. When properly designed, such experiments permit us to cast light upon and gain insight into questions that "ordinary" experiments cannot probe. As a case in point, we notice that what is "true" becomes fluid and mercurial. What is true for the participant is not necessarily true for the experimenter, or a third-party observer. This vacillating notion of "truth" pervades many aspects of modern research and how researchers work within the confines of the scientific paradigm to illuminate their hypotheses. As a magician turned scientist, I could sense that I was on the cusp of marrying these two worlds, with some interesting contributions to spare.[24]

*   *   *

When I was a kid, growing up in Israel in the 1970s and '80s, I learned of Mr. Uri Geller—a colorful and controversial compatriot who made waves all over the world by bending keys, spoons, and other pieces of silverware. Geller suggested, at times explicitly claimed, that he was doing it with his brain power: no tricks, just raw psychic energy. As a magically inclined youngster, I paid attention to his performances and statements, at least as much as I could glean from our black-and-white TV and daily newspapers.

Of course, I had the meaningful advantage of belonging to a network of magicians who knew better; however, I also noticed how many magic-naïve grown-ups all around me were slurping the Geller Kool-Aid with glimmering eyes. For me, it was an interesting demonstration of how people become enchanted when they think they are either watching a miracle, getting conned, or both—sometimes further elated by the notion that they are coming into contact with the supernatural, something that transcends science and surpasses reality.

I remember a specific after-dinner conversation my dad and I held in our living room. I was about fourteen, and my dad brought up Geller as an example of how little science knows about the paranormal, telepathy, and the power of the mind. That's when I first realized that my own dad, an electrical engineer, and a smart guy all around, seemed to have bought into the Geller narrative head over heels. I couldn't fathom how Dad, my inspiration and intellectual hero, would fall for something so preposterous.

For the better part of two hours, I tried to reason with him and appealed to his critical sense of thought that Geller cannot actually bend metal with his brain. I explained that he was a showman, and his performance only amounted to a magic trick. To prove my point, I even bent, Geller-like, several pieces of cutlery from our kitchen drawer using a few different methods. Dad was duly impressed for a few moments and inspected the mutilated silverware with surprise, but then he just smiled and dismissed the whole thing as "that's what you magicians do." It was a deflating experience but also an instructive one.

As a controversial entertainer, Geller showed that challenging people's relationship with the truth and planting doubt in their minds can ignite and fuel a form of wonder. Using his showmanship, charisma, and improvised interactions, he was able to perform metal-bending tricks in a way that riveted and elicited strong reactions from many spectators around the globe. Geller was able to sell the illusion of an alternative reality where he could defy physics and soar into the realm of metaphysics by using the power of his mind. "Any sufficiently advanced technology is indistinguishable from magic," science fiction author Arthur C. Clarke once wrote. In some ulterior way, perhaps Clarke captured how the mystifying charades and tormented cutlery championed by Geller heralded our current deep fakes—images, video clips, and soundtracks doctored by software to have anyone say or appear to do anything at all.

But formal magic and props are not always necessary. Sometimes just an unusual suggestion is enough to demonstrate a dramatic difference or illuminate a situation in an interesting way. As a postdoctoral fellow, I had the occasion to demonstrate how a psychology experiment beefed up with suggestion can give rise to some surprising results. The suggestion that I used was of the post-hypnotic kind—a suggestion given during a hypnosis session that lingers into common wakefulness even after the hypnotic session is over.

At the time, I was studying the Stroop interference effect, a curious psychological phenomenon whereby if I ask you to report the ink color of a printed word, you'd respond more slowly to incongruent words, such as "red" inked in green, than to congruent or neutral words, say "green" or "knife" inked in green, respectively. The interference from reading the word "red" slows down our ability to correctly respond with the ink color "green." This is happening, so goes the theory, because reading is an automatic, effortless process for proficient readers, and the spreading activation in the brain from reading "red" vigorously interferes with efficiently focusing, suppressing the associations with all things "red," and correctly selecting the response "green" for the ink color.

Since its discovery by Mr. Stroop in the 1930s, psychologists have been rigorously studying variations of this effect, which holds interesting implications for our scientific understanding concerning mechanisms of selective attention, executive function, and inhibitory control in humans. Thousands of papers and many academic dissertations have been published on this topic, and most cognitive scientists feel that they have a reasonable grasp of how this process comes about. Because the Stroop interference effect is so strong and inevitable, even if you know about it in advance, you'd still exhibit the trend in a carefully controlled experiment; few effects in psychology present with such a robust and ballistic resolve. I wanted to see whether a strong suggestion could crack, or at least make a dent in, the involuntary nature of the Stroop effect.

To achieve my goal, I invoked hypnosis and the power of a post-hypnotic suggestion. During hypnosis, I told dozens of participants— all native English speakers/readers and naïve to Stroop—that what they would see on the screen would be words in a foreign language that they don't understand, and that the symbols would feel like meaningless gobbledygook. Dividing my volunteers into two groups—highly and less-hypnotizable individuals—I had each person go through a standard Stroop task, once with the suggestion and once without it, in a counterbalanced order. At the end of the experiment, I let everyone know what happened.[25]

As it turned out, my suggestion was enough to dampen the Stroop interference effect in a dramatic way for the highly suggestible folks. It felt like magic. I was able to hack an allegedly unstoppable, automatic effect—deautomatize it, in a sense, by making people temporarily unable to read—and therefore had to think about the implications and the underlying mechanisms that were at work.

Was it possible that individuals who were more susceptible to a post-hypnotic suggestion could use their brains in a way so different from less hypnotizables? The answer seemed to be yes, and it took me a few more decades to experiment enough to prove it. It turns out, suggestion can have an intense effect on the cognitive processing and physiological response of highly suggestible people.

As it also happens, I later discovered that you don't even need to be highly suggestible to benefit from the impact of suggestion; everyone is agreeable to some suggestion. Moreover, when it comes to leveraging suggestions, magicians and scientists illuminate aspects of human behavior in surprising harmony. In fact, magicians can contribute to our understanding of the science of suggestion in ways that physicians, psychologists, cognitive neuroscientists, and social researchers cannot. In this regard, the concealed art of magic entails a profound dimension that nicely complements open-source science.[26]

Neuroenchantment

Because using placebos with patients may take advantage of their desperation and other vulnerabilities, IRBs find conducting deceptive research with healthy participants more acceptable.[27] With this fact in mind, one of my graduate students and I built a spiffy sham of a newly minted "brain machine." It featured assorted fragments of medical scrap centered around an old dome hair dryer, which she had collected from the sidewalk in front of her hair salon. I was already adept at making for-show physical apparatuses, so it didn't take us long to come up with a new "Spintronics Scanner." I knew that at least some participants would attempt to research and look up whatever technology we were using as part of our experimental scanner, so I had prepared labels and charts that suggested the involvement of spin electrons—a genuine field of physics.[28] After a thorough disinfection and a coat of paint, the meaningless assembly looked dauntingly medical. Following a long back-and-forth with the IRB, they finally gave their blessing, probably because we were running only healthy volunteers, and we were ready for our project to begin with the question: How much are individuals ready to believe when encountering improbable information under the guise of neuroscience? Phrased a different way, how enchanted are people by the overarching concept of "neuroscience" to explain away strange things?

Pointing to the hair-dryer dome, my graduate students told participants that brain scientists at the Montreal Neurological Institute had developed new experimental technology to decode resting-state brain activity to read the human mind. The idea that researchers can construe basic thought patterns from neural activity isn't new, but the dramatic presentation of the device and the prestigious reputation of the school, The Neuro at McGill University, surely enhanced the effect. As before, we embellished the new Spintronics machine with blinking lights, vibrating plates, sounds, dials, and screens displaying rotating three-dimensional brains. When entering the scanner room, warning signs similar to those found in Magnetic Resonance Imaging (MRI) rooms signaled their ominous messages about the presence of invisible magnetic fields, and all lab hands walked around wearing white coats.

Couched as an exercise in deciphering the neural correlates of thought, we asked participants to think of four items: a two-digit number, a three-digit number, a color, and a country. Participants chose freely and hid a note with their secret list in their pocket for later verification. The implied purpose of the study was to see whether with the Spintronics machine we would be able to interpret their brain dynamics to conclude what items they had chosen.

As a magician, I had plenty of secret ways to achieve this effect without resorting to Spintronics. I chose one simple method and as it turned out, almost everyone believed our Spintronics scam, including folks who should have known better.

I was especially keen on testing my undergraduate McGill psychology students. During that semester, I taught an advanced special topics course in which I made it clear that neuroscientists currently do NOT have the ability to perform the feat that this experiment purported to accomplish. To encourage students (and test their knowledge), the IRB allowed me to offer extra credit to students who opted to participate in several psychological experiments, including this one. As it turned out, a good number of students from my course showed up for the experiment. In class, these students understood that such technology was currently unavailable; but given the theatrical setup of the Spintronics scanner in the lab, this knowledge went out the window.

Around the same time, the annual conference of the Cognitive Neuroscience Society was taking place in Montréal, and the city bustled with cognitive neuroscientists from all over the world. I took advantage of this opportunity and invited some high-flying researchers (who knew me as a scientist, not a magician) to see our Spintronics setup. Many were skeptical, but some displayed the same trends I observed among the psychology students.

One person in particular, a prominent expert on medically induced comas, was curious about seeing my lab. "What's this Spintronics scanner?" Dr. Z asked with surprise when I gave him a tour of the lab.

"Oh, that's just an experimental pilot that one of my ambitious graduate students put together," I said casually, playing a role that I knew should lure him in deeper. "Like many before her, she also thinks that she can tease apart basic thought patterns from neural activity. I haven't yet had the time to look into her spin glass models and Spintronics technology, but her initial data look promising."

"She uses condensed matter physics and spin glass magnetic states?" Dr. Z exclaimed with eyes widening in shock, showing his physics insights.

"Well, she's far from an expert"—I treaded carefully with my professorial answer—"but these topics do come up in our lab meetings periodically."

"Let's give it a try," Dr. Z said with a facial expression dripping with eagerness. "I've never seen one of these scanners before."

"We can give it a try," I said with simulated reluctance, "but I don't usually run this experiment myself..."

When Dr. Z went through the motions of our Spintronics scam and the machine correctly identified his free choices of a two- and a three-digit number, a color, and a geographic location, his amazement was palpable. "This is unreal," he said, bedazzled. "I actually thought about these items in Hungarian, not even in English!"

"Well, that shouldn't really matter," I began to explain meekly, "I didn't tell you that..."

"I guess your scanner is able to capture the higher-level representations of my thoughts in some kind of universal, language-free space," Dr. Z cut me off with his academic musings, "where semantic concepts become network attractors..."

"It's actually simpler than that," I hushed him gently. "Please allow me to explain with the debriefing we provide at the end of this procedure..."

Dr. Z was a good sport. He understood that this study showed the power of set and setting but also something about the human tendency to relax critical thinking and overly trust brain-related images and data. I've coined a term for this concept—neuroenchantment.[29]

I was afraid word would start spreading around about the strange and magical studies that my lab was conducting, so I decided to dismantle the Spintronics assembly. But we had another mock scanner in the lab—one that we used to help people acclimate to the MRI environment prior to an actual scan. "Before we retire these sham contraptions entirely, do you think we can get one more bite at the apple?" I asked Jay, one of my graduate students and a fellow magician who knew how to pull off what I was about to unravel. Jay listened to my idea and got it immediately; he didn't need me to draw him a picture. He agreed on the spot, and one more time, we were off to the paranormal races.

Moving things farther afield, my team recruited sixty participants and led them to believe that our neuroimaging setup could both read and influence their thoughts.[30] In the Mind-Reading Task, the mock scanner appeared to guess numbers that the participants dreamed up; in the Mind-Influencing Task, the machine appeared to influence their choice of numbers in what we had dubbed "simulated thought insertion." In reality, of course, the entire effect was a theatrical piece performed by Magic Jay with his nonmagician graduate assistants.

"I expect that participants would feel less voluntary control over their decisions when they believe that the 'scanner' is influencing their choices," I shared with my research team before we started to collect data. "They may also make slower decisions in the Mind-Influencing

Task compared to the Mind-Reading Task." But I didn't anticipate some of the unusual reports that participants provided about their "thought insertion" experience.

Some claimed that the decision did not feel like it was their own: "It didn't really feel like I was making the choice," one said. "It kind of just happened." Another reported, "I kind of felt like [the number] came out of nowhere. So I felt like it...wasn't my choice. I don't know why I chose it." Others claimed they could not change the number: "I thought about trying to change [the number], but then...it doesn't." Another agreed: "I almost can't think of another one."

Some reported feeling that the number came from a source beyond their control. Sometimes the source was their own—apparently disobedient—brain. One claimed, "I was going...with thirty-four, but my brain just told me, 'No, that's not the number,' so I went... thirty-two, thirty-three, thirty-one, thirty." Another felt that "it's almost like my brain is shuffling through numbers until it...stuck to one." Some participants suggested the source was the machine: "I can't put my finger on it—it's just like...once the magnet turned on...I got four, and then I got seven..." For others, the source was unknown: "I felt like I was drawn to [the number]." Another said, "It really just felt like [the number] kind of came to me [from] somewhere else." One participant claimed: "I feel like it's a voice...dragging me from the number that already exists in my mind. I...feel some kind of force, or some kind of...image, or [something] trying to distract me from this number, and then I form [another] number."

Some participants also felt unusual sensations. One commented, "I pretty much felt like it was in my brain...I really noticed a kind of a pulsation...almost physical." Another noted, "I don't know why but my face feels really hot, like my head is really hot, but the rest of my body doesn't feel anything."

"Fascinating, isn't it?" I leaned over to Sam, a cognitive anthropologist on our team who got keenly interested in neuroenchantment. "Note how, in contrast, participants reported almost no unusual experiences or sensations in the Mind-Reading Task…"

"Absolutely!" Sam replied. "With this simulated thought insertion paradigm, we can influence feelings of voluntary control and may help model symptoms of mental disorders."

"That's a cool direction to explore," I agreed, "but I think we need to take it easy with all the magic on board. Plus, this mock scanner really needs to go because I need the space…"

Unfazed by these constraints, Sam already had an idea brewing for an experiment with children and teenagers.

"You might notice some unusual feelings, perhaps some tingling," Sam would say to the children as the motorized bed rolled them into the MRI tunnel. "Your brain is continuing to learn how to heal and to help you find this constant feeling of confidence."

This time the MRI machine was real, not a simulator. As it turned out, however, a mechanical problem forced the Brain Imaging Center to turn off its magnetic field. At the time, because restoring the magnet and resuming scans proved too costly, the Montreal Neurological Institute ended up with a defunct MRI device—a piece of furniture, really.[31] In this way, our young, unsuspecting participants believed that the machine was real, and they could not know it wasn't working. So, as far as they could tell, Sam at the console was able to glean information from their brains.

We decided to work with children not just because they are more suggestible than adults and often believe in magical thinking,[32] but because they are more "pristine" patients in the sense that their behavior contains fewer compensatory corrections than we would see in most adults.[33] Accordingly, again after IRB approval, we introduced pediatric patients to our "healing machine." We told them that our

MRI scanner could help them regain control over their symptoms. Moreover, we enlisted help from YouTube stars popular with kiddos to create a video promoting our experiment by explaining its benefits and describing the procedure in a language and vernacular appropriate for children. In this way, not only did our young patients immediately identify the stars and want to participate in our study, but they came with heightened motivation and curiosity.

In line with neuroenchantment, by peppering many "convincers"—for example, metal screening prior to entering the inactive MRI machine—we were able to assure the participating children that our paradigm was both viable and effective. Once we gained their trust, we were ready to influence their healing. In fact, we suspected that the suggestion would be so strong that we could even tell participants that any healing was due to their own brain rather than the machine. Indeed, we told them that the scanner used to take pictures of brains but was no longer functional, so we now used it as a healing apparatus. "But any healing," we emphasized to the participants, "would come from within your healing self." This OLP approach, loosely referred to as "open-label placebo," attempts to leverage the power of a placebo with the explicit communication that the participants are receiving one.[34] A long-term follow-up study on an OLP intervention showed long-lasting therapeutic effects five years later.[35]

We took several measures to heighten the neuroenhancement of medical magic. Props and accessories such as a fake nurse greeting the patients, a walk down a long, impressive hallway, and a ten-minute anticipation-building wait outside the lab were all part of our expectation building. Unlike the explicit deception we used in magic, here we told them mostly truths but relied on an implicit pretext, including the clinical hospital environment and our white coat attire, to enforce their assumptions and expectations.

Melissa was a thirteen-year-old girl who suffered from severe eczema and dermatillomania, a chronic skin-picking condition often related to anxiety. Melissa's mother, a nurse, would spend up to two hours a day treating her daughter's infected wounds, which covered her arms and parts of her face. Melissa explained to us, "It's like an addiction. I can't stop scratching, even when I sleep. I can't control it." Desperate for a solution, her mom learned about our experimental placebo research, and Melissa agreed to give it a shot. While in the scanner, Jay and Sam prompted her with multiple questions.

"As you go into the machine, you will relax more and more. Would you like to relax quickly or slowly today?" Sam asked.

"Uh, I can go quickly," Melissa responded with confidence. For the session to be successful, Sam planted the suggestion that the scanner would help Melissa heal.

Following Melissa's first session in the scanner, she and her mother reported that the skin picking had reduced drastically. The intervention is simple: the child goes into the scanner for roughly fifteen minutes, and we give them positive suggestions expressly tailored to help with their condition. Over 90 percent of parents reported improvements in the symptoms of their kids after their sessions. And many of these effects remained during a follow-up, a few weeks later.

After two more sessions, Melissa started feeling more relaxed, calm, and confident. She noticed that she wasn't picking as often and didn't have the urge to pick. After a while, her eczema started to heal. These preliminary results—preliminary as the sample was small—provided us with the hope that a placebo-based intervention may serve as an effective treatment option. Similar procedures, with the right framing, can support the notion that the mindset of patients plays a major role in some ailments and may facilitate healing processes.

To avoid any misconception about the trajectory of healing in these young children, it was imperative to ensure that for the children, this experiment wasn't successful solely due to our magical machine. My colleagues and I hypothesized that the children and teenagers have always had the ability to control their symptoms—they just needed to believe it themselves. In line with the way belief affects outcome, during the debriefing process, our research team bolstered the idea that their session would help them continue their healing processes.[36]

Sam asked Norman, a nine-year-old who suffered from attention deficit hyperactivity disorder (ADHD) and impulse control disorder, a question after his first session: "The amazing thing about the brain is that it has this fantastic power to heal itself, but now what we have been able to do here with the power of suggestion is to get your brain to work faster and better all the time. So how do you feel?"

"I want to go to sleep," Norman said almost immediately after being in the scanner for the first time. For a child with ADHD, Norman instantaneously appeared calmer. This outcome reminded me of the clinical effect we see when pairing a placebo with stimulant medication in children with ADHD,[37] as well as in other medical conditions.[38]

In addition to the debriefing session, participants were also given a specially programmed watch to remind the children between visits that the brain was healing itself. The idea that we gave them praise for their improvements was vital to our research mission and the long-term success of this procedure.

"Just having the watch with you will make you feel better. But remember, it's not the watch or the machine that's making you feel good, it's your brain." After their first session, patients were sent home and told to return to the lab in a few weeks for a progress check.

Niels, who suffered from migraine headaches, returned to the lab six weeks later and noted, "I haven't had a migraine at all. Not since the procedure. I'm actually really excited for high school." Melissa,

Norman, and Niels credit their improvements to the sham healing machine. They believed it was working for them and, admittedly, it did—this time without any sleight of hand. Our goal, of course, was to teach them that the power came from within them, not from the "scanner." This outlook is key. Indeed, mindset and narrative play a major role in how we construct treatments and situations.[39]

These results are encouraging and a compelling sign of how powerful suggestions and our willingness to believe in them can deliver, especially when it comes to brain stimulation.[40] However, they are still anecdotal testimonials of the type that is difficult to generalize because they can be unreliable.

Luckily, we have a method to ascertain reliability that is one of the most important contributions of humans—science. Today, we often resort to metascience—research on research—a field that examines and combines all research on a scientific question using a wide variety of methodologies. What happens when we apply science and metascience to demonstrate whether the promising results of suggestion hold up when we control for and compare all the right conditions? We can begin to tease apart effects that are statistically significant, or in line with our hypothesis, from those that are clinically significant— those that hold meaning in reality.

CHAPTER 4

Suggestion Changes Your Physiology

I magine that I place an ordinary milkshake in front of you. It's a typical dessert, thick with sugar and cream and all the other rich ingredients you'd expect in a shake. You inspect its label; "Indulgent" it reads, with over six hundred calories. Already, you feel your mouth beginning to water, your taste buds anticipating the sumptuous flavor. I ask you to drink until you feel full. How much of the shake do you think you'd gulp down?

Now, let's reset the stage. I hand you a different plain vanilla milkshake, nearly matching the one in the previous scenario. Only on this one, the label reads "Sensible." Evidently, it's the newest prototype from a state-of-the-art line of diet, low-fat desserts. Only 140 calories. I ask you the same question: How much of the shake would you drink?

Most health-minded people would consume less of the high-calorie milkshake. This is exactly what happened in an experiment at Stanford University. Participants who drank the richer 620-calorie dessert experienced the effects of a sharp decline in ghrelin, the "hunger hormone"—a high level of ghrelin increases appetite, and a low level induces fullness. Meanwhile, ghrelin levels in those who drank the

140-calorie shake remained unchanged. These results should present little surprise, except both shakes were the same—identical![1]

Same taste, same size, same calorie count—the only difference was the suggestion on the label. Although participants felt as if they were drinking two different shakes, the suggestion shaped their reality. Not only that, but their caloric expectations physically altered their metabolism, allowing them to consume less yet feel more full. These simple findings join a wealth of similar data with immense implications.[2]

For example, you'd think that these collective results would finally do away with the "calories in/calories out" model that governs the literature on medical dieting. Burning more calories each day than you consume may have been the dietary advice you heard in the past, but it doesn't work for everyone.[3] Counting calories alone doesn't work because, by itself, decreasing calorie intake has but a limited short-term influence.[4] But if you realize that "all calories are not created equal," especially if they are not perceived equally, you can change the game. The checkered history of nutritional science taught us that the calorie-counting focus got us into trouble with low-fat diets in the first place; we now know that where calories come from also matters. Perhaps next we need to pay more attention to our mindsets and suggestions about food.

When I broach the relationship between expectations and physiological responses, most people naturally think about placebos:[5] physiological changes occurring in the body without specific biological stimulation.[6] For example, when Harvard researchers asked study participants to role-play air force pilots—in other words, play the part of professionals with excellent eyesight—they displayed better vision than control participants.[7] (Please don't throw away your eyeglasses just yet; wait until you read the whole book.)

Throughout my career, I have been repeatedly surprised to discover how the role of psychological states remains largely undervalued.[8] This fact is bewildering if you consider the broad range of data describing mental influence over the body. To be sure, the connection between mind and body has received limited attention in the study of metabolic disorders, such as diabetes, for example.

Most folks who grow older, heavier, and more sedentary learn about a chronic disease that affects millions: type 2 diabetes, the most common form of diabetes.[9] As a kid, I experienced this disease vicariously through my grandmother. Her symptoms included periodic rises in blood glucose levels because her body produced insufficient insulin and resisted the effects of insulin. Subsequently, she occasionally suffered from short-term severe shock and later developed multiple long-term complications: from kidney disease and vision problems to neuropathies and strokes.[10] We now know that while genetic factors present a strong biological factor in type 2 diabetes,[11] obesity poses a key environmental trigger.[12]

Glucose levels in people with type 2 diabetes change according to a specific time course. My grandmother used to prick her fingers multiple times a day to draw a drop of blood and keep track of her sugar levels, but today, we have technology that can continuously monitor glucose levels over weeks. Most medical experts assume that physiological factors alone govern how glucose measures wax and wane over time,[13] but psychological factors and environmental cues likely play a role as well. For example, haven't you ever gotten hungry—without feeling the slightest bit peckish just moments before—when you noticed it was lunchtime or upon catching a waft or glimpse of an attractive food?[14] I certainly have.

Against this backdrop, researchers have reported a connection between changing the perception of time in individuals with type 2 diabetes and their blood glucose levels: the level of sugar in their blood changed in accordance with how much time participants *believed* had passed, rather than how much time had actually passed.[15] Another study that used identical ingredients but deceptive nutrition facts labels showed how, against common wisdom, suggestion influenced the physiological parameter of glucose level.[16] Blood glucose levels increased when individuals diagnosed with type 2 diabetes falsely believed that their beverage had high sugar content, as seen on a made-up label. Again, these findings join a line of pressing evidence showing that the use of suggestion and mindset can influence the management of diabetes,[17] yet current programs to manage type 2 diabetes draw on diet, exercise, and medications only. My grandmother did not have a chance to benefit from these leads; perhaps our generation, and future ones, will.

We don't need milkshakes, time-warped diabetics, or phony nutritional labels to understand that our perception creates our reality—and to appreciate how it relates to suggestibility. Virtually all we need is a plump, juicy lemon—or the idea of one. Hold it in front of you, and notice the unmistakable smell of this yellow citrus fruit as you cut it in half. Now tilt your head and elevate one half above your head and slowly squeeze it into your mouth, so that a good trickle of acidic juice drops into contact with your taste buds. Feel how your mouth waters in dreadful anticipation of the sting that accompanies this tangy liquid. That's how suggestion shapes physiology.

The Magician-Scientist

I trust that you have had occasion to fool your doctor, perhaps as a kid keen on avoiding school; but has your doctor ever fooled you? Back when I was taking a clinical elective in behavioral neurology, the teacher held one session in the epilepsy unit of the teaching hospital. In a white coat, the attending neurologist, Dr. N, presided over the session. He was about to demonstrate a deceptive "provocation test" on a young female patient, Vero, who complained about debilitating seizures. Vero described bouts of uncontrolled shaking movements involving much of her body and even showed us a striking homemade video of some of her symptoms, captured by a relative. For all kinds of medical reasons, the epilepsy team suspected that her seizures were psychological—in other words, nonepileptic—rather than neurological (epileptic). Dr. N decided to use a challenge test to confirm this diagnosis. Vero was unaware of the looming procedure.

To begin, Dr. N attached some electrodes to her scalp in order to record her brain electrical activity using electroencephalography (EEG) and hooked up a cannula (intravenous line) to her arm, telling her that she was receiving a solution designed to provoke a seizure. However, although the electrodes measured genuine EEG, the tube in her arm ran plain saline—a benign solution that wouldn't trigger a seizure. If Vero developed a seizure, Dr. N would stop the infusion and help Vero recover. The lack of abnormal EEG activity from Vero during such an episode would prove the diagnosis of a nonepileptic—psychologically driven—seizure. It was an exercise in theatrical medicine.

It all sounded super interesting, but before long, a small problem became apparent. Although Dr. N was a respected clinical neurologist and a senior gentleman, he was no great actor and displayed a rather stiff bedside manner. Throughout his entire attempt at scene-making, Vero just sat there, seemingly bored. Although Dr. N was going through the motions of the provocation test, we—the students in attendance—all felt it was a rather clumsy, contrived effort to propel Vero into developing a nonepileptic seizure. After what felt like long minutes of failed attempts by Dr. N, he finally gave up, and the vibe in the room turned awkward. The magician-hypnotist in me felt compelled to spring into action.

"But look here, Professor," I reached for the saline bag and started futzing around with it, "the valve had remained shut! Here, I opened it now and Vero will seize very shortly." Stunned by my bold move, Dr. N froze, speechless. But Vero perked up; this development got her attention. She watched me closely as I immediately leaned over the screen, which showed the live EEG traces from her brain. I pointed excitedly to one of the squiggles on the display and said loudly, "Here, you can already see the seizure brewing. It's gathering momentum and will soon spread all over. Make way for her, she may need some room!" Within twenty seconds, Vero was convulsing on the floor.

Theatrical antics and contextual suggestions are part and parcel of many clinical interventions and medical healing. They unmistakably affect physiology and influence outcomes. In epilepsy, for example, more than 20 percent of patients who show up in hospitals with seizures or get treatment for epileptic seizures likely have nonepileptic seizures; 10–25 percent have both. Whereas medications tend to manage neurological epilepsy, they are less effective for the seemingly psychological nonepileptic seizures.

Although we were able to help Vero—she is doing well today—a few lingering questions remain: Do the minimal physical risks associated with a saline infusion and the definitive diagnosis of nonepileptic seizures justify lying to Vero, albeit to spare her from taking daily anticonvulsant medication? And should we tell Vero about our "little ruse" after everything settles down? After all, such a discovery may lead her to lose confidence in other doctors, perhaps in the entire field of medicine, and may therefore present obstacles to her future medical care.

With these serious ethical considerations, it's clear why deceptive provocation tests have mainly been abandoned in some settings (yet still thrive in others). But is there a formulation that mixes magic-like elements and science into a judicious and ethical elixir that could benefit patients? Can we use theatrics to influence expectations and enhance healing? Are there critical scientific questions that a sprinkle of magic can help further investigate? You bet there are.

Simon Says

When we first met, Simon was a quiet young boy, about twelve years old. He sat in a chair across from my desk, hands in his lap, politely explaining his condition.

"It feels like someone else is doing it," he mumbled.

"So you can't really control it?" I asked, although I already knew the answer.

"Sometimes I can. It's a bit like sneezing or scratching an itch. Sometimes I can hold it in, but most of the time I just have to let it out. I can only fight it for so long..." he answered with a bit more vigor. "And if I hold back, I'll need to tic more later."

I was impressed with how simply and eloquently this preteen could describe his complex "rebound" behavior.

At age ten, Simon was diagnosed with Tourette's syndrome (TS). As we spoke, his body would twist and jolt with sudden, explosive tics, including head jerking, eye blinking, and facial contortions. And every now and then, he would also utter an unexpected vocalization—a muffled bark of sorts. The juxtaposition of his normally calm composure against these abrupt tics was striking.

Simon and his mother had consulted a long line of experts before finding their way to me. They'd seen child psychiatrists, pediatric neurologists, clinical psychologists, and movement disorder specialists.

He'd been on and off multiple medications. The list was long: from antipsychotics (both conventional and atypical) to a whole assortment of neuroleptics and dopamine antagonists, alpha-adrenergic receptor agonists, antidepressants, cannabinoids, and even injections of botulinum toxin to his affected muscles. These interventions either helped too little or came with side effects—including tremendous weight gain and extreme cognitive dulling—as devastating as the symptoms they were intended to relieve.

He also tried many types of behavioral therapy—including operant conditioning, awareness training, habit reversal, anxiety management, and massed negative practice—and several diets, as well as complementary and alternative interventions. Still, his tics wouldn't budge. These less-conventional options fell flat as well. Simon and his family were at the end of their rope, desperate for better treatment choices.

I knew that the symptoms of TS might arise from anatomical and functional disturbances in the Cortico-Striatal-Thalamo-Cortical (CSTC) circuits in Simon's brain. This neural "freeway loop" connects parts of our cortex, the seat of our higher brain functions, with more deeply embedded, subcortical striatum—the nerves that control motor and action planning—and also cognition and affective experience, decision-making, motivation, reinforcement, and reward. From the striatum, this pathway continues to the thalamus—an intricate relay station that processes both motor and sensory signals—which then loops back to the cortex.[18]

Moreover, using functional imaging technologies of the living human brain, my colleagues and I have been able to unravel some of the complex interactions among CSTC components and show how teasing apart this neural dynamic may inhibit tics.[19] Because TS often presents alongside attention deficit hyperactivity disorder, obsessive-compulsive disorder, anxiety, depression, and self-injurious behaviors, we dreamed up and experimented with a few creative hypotheses. The main idea:

perhaps we can leverage suggestion to regulate CSTC function and thereby help the symptoms of individuals with TS.[20]

* * *

Putting on my white lab coat, I hunched down to Simon's height and quickly looked around the room as if I were about to share confidential information. "Hey, Simon," I whispered, "can I trust you to keep a secret?"—a line I often used back in my stage magic days.

Simon perked up, his curiosity piqued. He looked at me intently, nodded his head, and mouthed a dawdling "Y-yes."

"Have I ever shown you my Tic Detector machine?"

"No!" he exclaimed with a smile.

"*Shhhhhh,*" I hushed him softly. "We don't want others to know about this machine, okay? It's still under development. But I can show it to you if you promise not to tell anyone."

I could see that Simon felt duly special, and as a function of his increased engagement, his tics were subsiding.

"Follow me," I mimed. "I will show you."

Simon quietly trailed me to the next room in the laboratory. I paused to look around periodically, pretending to make sure that no one was following us.

Pulling off the covers in one downward motion, I unveiled a large, medical-looking device with a prominent lens on a formidable tripod. "Wow!" Simon huffed at the unusual sight, riding on the tail end of one of his involuntary vocal shrieks, "That thing looks weird."

"It does, doesn't it?" I reached toward the main power switch. "Should I turn it on?"

"What does it do?" he asked hesitantly.

"It's a Tic Detector," I said with feigned surprise. "It detects tics. Don't you remember our conversation next door?"

"Yeah...I mean, how does it work? Does it hurt?"

"Not one bit! It works like magic. Let me show you..." I flipped the control switch on.

Simon followed the array of lights and dials. "It's booting up and calibrating," I said as I peered over the multiple knobs. "Let's wait until it completes its internal check." The machine made a long beeping sound. "I think it's ready."

"If you go sit over there"—I gestured toward a chair in the middle of the room—"I can point the scan beam at you, and the Tic Detector will beep every time it identifies a tic coming..."

Simon sat in the chair, still processing the whole scenario. I pointed the lens at him and stepped off to the side. "Here we go!"

About twenty tic-free seconds later, a different, sharper beep came from the machine. Simon jumped in his seat, startled. "The machine detected a tic forming," I explained. "Can you feel one brewing inside you?"

Simon nodded slowly with a sense of wonder.

"You see, it's working!" I continued. "Are you feeling this tic forming because it's really brewing or because we're talking about it?" I asked.

Simon seemed confused by the question. Before he could answer, I continued, "You know, this machine is so powerful that it can detect tics you don't even notice—early tics, baby tics, etc. But if you give it a bit of time, you will feel these tiny tics build up and mature into a tic you can feel."

*　*　*

Simon didn't know that this "Tic Detector" was an assortment of broken electronic devices I had collected from the graveyard section of the hospital basement, where all the old medical instrumentation ended up before going into the dumpster. The centerpiece of this assembly of junk featured a lofty TV camera used in the audiovisual studios back in the 1970s. I adorned it with a mishmash of dials, buttons, wires, and needles and added some touches of old microchips and dusty transistors to form a meaningless heap of electronic garbage—and *voila*, the Tic Detector.

The only working piece of equipment was a sound-generating box; I set it to randomly beep every thirty seconds or so. The rest was all showmanship—a magic trick of sorts. I wanted Simon to believe that the Tic Detector was able to detect even the most incipient tics, those so nascent that they were still unavailable to him. It was the old ruse: "Either I win, or you lose." If the machine beeped and Simon said he felt a tic coming, it's a hit! If he didn't feel it coming, I'd explain that the machine may be more sensitive than he is.

"Just wait a bit and you'll feel it…" I reassured.

If there's one thing you need to know about TS it's this: whether a cheep or a deep beep, if you wait long enough, a tic will surface—it's a hit, no matter what Simon says!

* * *

Simon and I spent about twenty minutes playing with the Tic Detector. It was obvious that he bought into this charade—the whole kit and caboodle—because every time the machine beeped, he would subsequently tic. Now that he believed I could detect his tics, I needed to rid him of them.

"Would you like to see the Tic Deflector now, Simon?" I asked him, pointing to another covered object in the corner of the room.

"What's that?" Simon replied with surprise.

"It's a different machine that can prevent and melt away the tics that the Tic Detector identifies." I explained. "Let me show you."

It took one strong yank on the tarp to reveal another pile of discarded medical scraps. Only the Tic Deflector made a "bop" sound, rather than a "beep."

"Before I turn it on," I whispered to Simon, "you should connect the optical cable so that the Tic Detector can communicate with the Tic Deflector. The connecting cable runs right under your chair..."

Simon followed my instructions, and with his "help," we were able to join the two machines together. I switched on the Tic Deflector, and we both watched it with anticipation, waiting for the calibration cycle to complete. "Once it receives all the information from the Tic Detector, the Tic Deflector disarms the tic and erases it. Once gone, you don't need to tic it out," I explained to Simon, who stared wide-eyed at the setup.

Once again, Simon sat there, only this time tic-free and hardly moving, with strings of auditory beeps and bops flying overhead. He stayed like that for about thirty minutes, and I invited Mom in to see. She looked at the scene in amazement. "What did you give him?" she asked with a hint of unease, likely scarred by all the different medications and side effects they had to endure over the past years.

My answer was honest yet profound: "Just a story and some rickety pieces of scrap..."

* * *

Weeks after the Tic Detector "experiment" with Simon, when I presented my observations to some of my departmental colleagues, I expected them to embrace the clinical merits of suggestion. After all, my demonstration with Simon showed how we could possibly use suggestion to help individuals with TS.

But things didn't go as planned.

"Theatrical medicine?" one of the senior psychiatrists cried with consternation. "That's a BIG ethical no-no, you know..."

"Well, yes," I responded with deference, "but Simon had already tried everything that conventional medicine could offer, and nothing worked. Plus, the Tic Detector seemed to help without all the harmful side effects of medication. In line with the Declaration of Helsinki governing ethical principles for medical research with humans,[21] is it not our duty to look into it more? Reading Article 37 of this declaration, isn't it less ethical to *not* help these people?"

World Medical Association Declaration of Helsinki

Ethical Principles for Medical Research Involving Human Subjects

Unproven Interventions in Clinical Practice

37. In the treatment of an individual patient, where proven interventions do not exist or other known interventions have been ineffective, the physician, after seeking expert advice, with informed consent from the patient or a legally authorised representative, may use an unproven intervention if in the physician's judgement it offers hope of saving life, re-establishing health or alleviating suffering. This intervention should subsequently be made the object of research, designed to evaluate its safety and efficacy. In all cases, new information must be recorded and, where appropriate, made publicly available.

"Don't you know that people with TS can show fewer tics when they play an engaging video game, pay attention to a captivating task, or during sleep?" the bespectacled professor asked with a mocking tone.

"Of course, I'm aware," I assured him, "but when these individuals are under the impression that a medical machine can both discover and obliterate their tics, they don't need to twitch them out. There's no 'rebound' that follows thereafter. They seem to lose not just their motor 'premonitions,' but also ownership over their own tics. That's a completely different narrative, entrenched in top-down control, and it seems to make a clinical difference—"

"You want to treat Tourette's with a narrative?" he scoffed.

"Look, the prevalence of tic disorders in the classroom is between 5 to 20 percent of children, with impairments occurring in one to ten per one thousand school-aged children worldwide,"[22] I paused. "It's important to help these kids in any way we can. The 'Tic Detector' suggestion is simple, effective, and safe."

His face reddened: "Your deception is breaking our system!"

"I'm not your adversary," I replied in my best calming voice, "I am your colleague, but if I break, it's in order to reveal, cure, and heal."

The psychiatrist shook his head and walked off in a huff.

I didn't run after him. His position was familiar to me. Although parents and caregivers understandably lie or otherwise find creative ways to influence their children with all sorts of imaginative persuasions, mixing *clinical* care with deception is a slippery slope.[23] Moreover, case studies entrenched in testimonials, small sample sizes, and anecdotal reports may indicate that an untested treatment is promising—even compelling—and worthy of formal scientific study, but they hardly serve as rigorous clinical evidence. There is a complexity here. On the one hand, many agents—from psychologists and therapists to teachers and parents—practice suggestion liberally. On the other hand, modern psychiatrists seldom focus their clinical interventions on the power of narrative, suggestion, or magical maneuvers grounded in top-down control. Common wisdom may even deem such clinical practice as "bad form," and physicians think that they ought not to engage their patients in this way lest they lose their medical license.[24]

Similar to the conceptual separation between "psychology" and "psychiatry," many people may share the impression that scientists work relentlessly to uncover the truth, whereas magicians go out of their way to deceive. To put it more bluntly, scientists are "honest, systematic, and virtuous," while magicians are "dishonest, sleazy, and sly." In the case of Simon, these two worlds seem to collide. The dual nature of a magician-scientist reeks of Dr. Jekyll and Mr. Hyde—scientist: instinctively good and ethical, a thoughtful intellectual; magician: inherently deceptive and theatrical, a slick performer.

And yet, Simon got off medication and was doing so much better. Of the about half dozen children and teenagers diagnosed with TS, just like Simon, that I have seen over the years across three universities, about 85 percent responded well immediately following such interventions. They would further show a marked reduction in tics for weeks and sometimes months thereafter, especially if I had a way to periodically boost or reinforce these effects. Without a formal study, however, these observations are just clinical vignettes about the potential power of suggestion.

In line with Article 37 from the Declaration of Helsinki, this situation presents a conundrum: Should we not systematically study these phenomena to gain a better scientific understanding of the underlying mechanisms, and find ethical and judicious ways to leverage suggestion to benefit those individuals who need them the most? I'd like to say yes, but there are nuanced challenges to taking these clinical leads and turning them into a proper scientific trajectory.

The Power of Suggestion—Warts and All

Suggestions can change your life, but perhaps in a way that is different from what you have been told by advertisers,[25] architects,[26] New

Age gurus,[27] and personal development books.[28] As a scientist, I cannot endorse pseudoscientific ideas that purport to explain how "positive energy" attracts "positive things" into your life, governs your thinking and actions, and empowers you to achieve anything you can imagine.[29] Moreover, as a magician, I cannot ratify self-help books that promote a singular concept as a magical panacea.[30] However, I can certainly provide practical take-home messages that are entrenched in our current understanding and knowledge of the science of suggestion, as nascent and incomplete as this field appears at this time.

Suggestion was all the rage when the tic-detecting machine affected Simon, but the transition from working with clinical patients, such as Simon, to experimenting with healthy volunteers is sometimes tricky because the shift from clinical to healthy populations falls within a regulatory twilight zone. For example, consider common viral warts (verruca vulgaris), one of the most common diseases of the skin. With a wide range of available treatments, warts may appear in any area of the skin, but they mostly affect the hands and feet.[31] Because researchers can assess wart loss both easily and objectively, warts have been the most popularly studied dermatological disorder with respect to psychological interventions.[32]

Nearly a century ago, suggestion became one of the most popular forms of psychological intervention for inducing wart loss.[33] One early study examined whether the introduction of a sham "wart-killing machine" would successfully help people get rid of warts. In 1927, Bruno Bloch reported how he blindfolded patients with warts and placed their infected hands into an "electric machine." Unbeknownst to them, his apparatus made a humming sound but little more. To further heighten the effects of suggestion, Bloch also painted their warts with methylene blue, an innocuous pigment. After removing their blindfold, Bloch instructed his patients not to touch their warts until the blue color had disappeared.[34]

Of the 179 patients who were available for a three-month follow-up, 78.5 percent were wart-free—about 30 percent lost their warts within two weeks after the treatment session, an additional 43 percent showed complete remission within one month, and a further 18 percent lost their warts within three months of treatment. Altogether, Bloch reported a success rate of over 50 percent one month after treatment. Warts typically go away on their own within two years. Remarkably, Bloch claimed that explaining to his more educated patients that they were, in fact, being treated by suggestion did not reduce success rates.[35] In many ways, Bloch further propelled not just the idea of machine-based sham interventions but also our modern discussion of open-label placebos—inert agents taken without deception wherein participants explicitly know that they are receiving a placebo.[36] Top it up with controversial sham surgeries and the surprising clinical and ethical effectiveness of taking imaginary pills[37]—yes, this is not a typo—and you can appreciate the centrality of suggestion for humans.

Bloch recruited participants with warts, so they were not "healthy volunteers." Still, perhaps, for all kinds of practical purposes, they were closer to healthy volunteers than, say, individuals diagnosed with TS. In some oblique way, this type of study—although I was completely unaware of it at the time of my informal sham-machine explorations—may have indirectly inspired preliminary Simon-like efforts.

Nicholas P. Spanos, a creative social psychologist who liked to challenge common beliefs, including about hypnosis, is relevant here. In a string of multiple experiments throughout his productive career and leading up to his untimely passing in a plane crash that he piloted, Spanos and his colleagues claimed that "hypnosis" refers to suggested behaviors that a participant chooses whether or not to comply with. Together with his research team, he compared a Bloch-like wart-killing machine (placebo), topical salicylic acid, or no treatment control with hypnotic suggestion. A six-week follow-up showed that hypnosis was the most effective in reducing warts.[38]

Over the years, I have spoken with many seasoned skin doctors, as well as dermatology residents, about effective treatments for warts. They all agreed that psychological elements such as expectations, stress, mindset, and anxiety could affect the appearance and disappearance of warts and other skin conditions. They readily acknowledged that certain aspects of dermatology, including warts, responded well to psychological interventions. And yet almost every single one of them shied away from recommending such interventions and always preferred prescribing a topical medication (e.g., salicylic acid) or another intervention (e.g., cryotherapy), even when these approaches proved repeatedly unsuccessful for specific patients. In some places—for example, North America—physicians seem especially uncomfortable recommending psychological interventions and suggestion-based treatment plans, perhaps due to evidence-based hesitations, fear of malpractice, and potential concerns about regulatory oversight of their medical licensure.

This modus vivendi was understandable when research into the underlying mechanisms that mediate wart loss was essentially nonexistent because we had little understanding of the physiological processes that facilitated the relationship between psychological treatment and the eventual loss of warts. But our understanding of how the brain and immune system interact has changed substantially since. Whereas initially we thought that the brain was immune-privileged and isolated from the rest of the body, we now have a better scientific understanding of the relationship between brain states and immune response.[39]

*   *   *

We are beginning to uncover how mindset and mental states can have a great impact on how sick we get and how quickly and completely we recover by considering brain-immunity interactions. Today, we are beginning to harness suggestions and their effects on the brain's control over the body's immune responses[40]—for boosting the placebo effect,[41] hindering a spectrum of diseases from autoimmunity to cancer,[42] enhancing vaccination response,[43] and even reevaluating functional disorders that have long been shunned as "psychological."[44] These insights pave the road to a more scientific understanding of how we can better treat psychosomatic—or mind-body—conditions.[45]

As a case in point, decades of research have documented the contribution of our psychological outlook to cardiovascular health.[46] Moreover, we now know that a positive mindset can lead to better outcomes in many chronic conditions, although the underlying mechanisms remain largely elusive.[47] But now scientists are beginning to unravel why stimulating a region of the brain involved in positive emotion and motivation can influence how the heart heals.

We already know that, at least in mice, activating the brain reward center—the ventral tegmental area (VTA)—appears to instigate immune changes that can fight cancer: activation of the VTA restrains cells in the bone marrow that usually inhibit immune response, thereby priming the immune system to fight the cancer.[48] But now we are beginning to see that stimulating VTA neurons also contributes to the reduction of cardiac scar tissue. This insight complements another: stimulating neurons in the mouse hypothalamus—a regulator of body temperature, hunger and thirst, mood, libido, blood pressure, and sleep—can trigger an immune response[49] that may even control the growth of tumors.

Moreover, neural cells in the insula—deep within the brain—store memories of past flares of gut inflammation; stimulating those neurons of the insula relaunches the immune response.[50] In other words, such immune reactions could start up even in the absence of a gut trigger.

This situation may apply to common psychosomatic conditions such as irritable bowel syndrome (IBS), which presents with symptoms of abdominal pain alongside diarrhea or constipation—or both—but without any pathological finding in the digestive tract. This mechanism explains why suggestions[51] and open-label placebos can work well in cases such as IBS.[52]

The symptoms of IBS often get worse as a function of negative psychological states. Just like a positive mindset can be helpful, a negative mental disposition can compromise the immune response by dampening the neural activity of specific brain circuits that potentiate immune cells during acute stress.[53] Such findings inspire researchers to look for creative ways to influence the stress levels of participants while measuring the subsequent changes in the immune response. In this way, the link between psychology and physiology is becoming more palpable. Also, we are getting closer to dropping the prefix "psycho" from the term "psychosomatic," ostensibly moving us nearer to the "somatic."[54]

We now know that the brain performs a form of "immune surveillance" and draws on a complex brain-immune network that involves multiple peripheral immunological players.[55] This brain-immune relationship plays a role across development and into aging, as well as during various illnesses.[56] Moreover, leveraging our insights about this brain-immunity interface may reveal new therapeutic targets for several neurological disorders.[57]

CHAPTER 5

Can Clinical Suggestions Heal and Cure?

Is Honesty the Best Policy?

If you don't read the sign, it's hard to know that building eighty-three at 25 Sigmund-Freud-Straße is the location of the Department of Epileptology at the University Hospital of Bonn in Germany. The people who come to this building either suffer from or know how to identify and treat abnormal electrical discharges in the brain that often come with epilepsy. In clinical terms, epilepsy refers to (a) two or more unprovoked seizures occurring more than twenty-four hours apart, or (b) one unprovoked seizure and a 60 percent or greater probability that a second seizure will occur during the next ten years. The main treatment options for epilepsy draw on a combination of antiepileptic medications, neurosurgery to remove the problem area in the brain that's causing the seizures (if it's small enough and easily accessible), implanting a tiny electrical device to detect and counter developing seizures, special diets, and specific psychedelic substances such as medical marijuana.[1]

But in this three-story, twenty-eight-bed clinic, which fuses straight, blocked edges with modern architectural components, another unusual opportunity awaits to explore seizures: suggestible neurophysiology. In this building—one of the premier epilepsy centers in Europe, and arguably in the world—in addition to standard neurology, the medical crew also employs the more controversial technique of provocation tests. As part of medical diagnosis, the clinicians may expose some patients to a sham substance, which typically provokes no response, while explicitly suggesting to them that it will trigger a seizure.

The use of provocation tests almost mandates that some form of deceit creeps into the medical procedure. For obvious reasons and as we saw in Chapter 2, ethicists typically frown on this practice, and it has been largely abandoned in the more litigious US.[2] Whereas most clinicians shy away from provocation (sometimes called "challenge") tests because of ethical concerns, the Bonn hospital still maintains one of the last bastions of challenge tests in clinical epilepsy. Under this roof, neurologists, neuropsychologists, clinicians, and researchers acknowledge the ethical caveats and vehement disapprovals associated with seizure provocation, yet are courageous enough to stand up to criticism and balance these concerns with the clinical value, medical insight, and scientific understanding afforded by such challenge tests.[3] These practitioners and researchers are top epilepsy experts. They aren't agent provocateurs; instead, they are medical specialists who judiciously use provocation tests in the clinic.

The vast majority of patients who come through this door either have been diagnosed with epilepsy or show epileptic symptoms. Outpatients come and go, populating the ground floor. The inpatient ward on the first floor of the building houses ten monitoring rooms for conducting video EEG—affixing electrodes on the scalp of a patient to continuously record their brain wave activity while simultaneously collecting video footage of their behavior. With this setup, clinicians can capture the underlying neural footprints and associated behavior leading to a seizure when it occurs. In most cases, such sudden attacks or electric storms transpire unexpectedly, without any prior warning. Accordingly, the video EEG sessions may sometimes go on for days before a seizure surfaces.

Catching a seizure on video, alongside its underlying EEG, provides extremely useful clinical information.[4] It allows experts to bind the anatomical source of the abnormality in the brain to the behavior that accompanies it. One of the main problems is the wait; we can twiddle our thumbs over many hours before an eruption will show on the video-EEG monitor. Whereas European epileptologists use video EEG for a day, two, or maybe three, their American colleagues sometimes opt for a week or longer. Many reasons—financial, legal, cultural (private versus socialized medical systems), but perhaps not strictly medical—likely account for this difference.[5]

To shorten the amount of time required by video EEG to document a seizure, most epilepsy centers rely on assorted techniques[6]—for example, making patients go through sleep deprivation, hyperventilation, exposure to flashing lights and images, repetitive movements, and withdrawal of neuroleptic medication—to shorten the monitoring session.[7] These "tricks" create the "ideal" welcoming environment for a paroxysm to surface.

However, not all seizures are the same; for example, some are epileptic and some are nonepileptic. We have already discussed how suggestion can invoke a seizure, but when such behavior comes about because of a suggestion, clinicians label the seizure as "psychogenic" (i.e., originating from a psychological cause) and nonepileptic. This distinction is a bit tenuous, not just because patients can suffer from both epileptic and nonepileptic seizures, but also because the separation between the psychosocial and physiological domains is hardly cut-and-dried.

Most people, certainly most physicians, receive training that favors a bottom-up view of physiology: genes dictate the translation of proteins and carry their effects all the way up the hierarchy to governing behavior. But top-down influences also matter.[8] Things like expectations, cultural affordances,[9] social norms, and the interaction between the experimenter and the experimental participant—say in lab studies—also govern our physiology.[10]

When neuroscientists study cognitive function, they consider different factors, including anatomy, evolution, and patterns of control and governance. However, as with most metaphors that illustrate how humans manifest thought and action, these models often border on the philosophical. The take-home message is simple: whereas bottom-up emphasizes sensory information that comes from outside the brain, top-down acts on information that comes from ideas and expectations within the brain. Suggestion plays a key role in establishing, maintaining, and fueling top-down influences.

In addition to bottom-up effects from genes and sensory input, top-down effects come from sociocultural influences, community values, psychological expectations, and ethical mores that shape our consciousness. Having a suggestible brain equipped with top-down capabilities gives us an evolutionary advantage and guides our ability to learn, adapt, and thrive in line with a fast-changing culture. And yet many scientists have a bias or preference for bottom-up over top-down. For example, a bottom-up epileptic seizure is neither more nor less real than a top-down psychogenic seizure invoked by suggestion. However, individuals diagnosed with a psychogenic label often end up in a "treatment void" with few therapeutic options—neurologists and psychiatrists seldom offer effective care for them.[11]

At this Bonn-based clinic, practitioners have refined provocation tests—a controversial clinical procedure complete with placebo infusions, psychological suggestions, and expectation-building steps—to discern an epileptic fit from a nonepileptic one (as we learned in Chapter 4). The goal of this unusual approach is to offer patients suffering from psychogenic seizures viable treatment options, which are so sorely lacking, rather than let them drop into a clinical dust heap. In this German clinic, practitioners are looking for, and experiment with, effective ways to offer treatment for psychogenic nonepileptic seizures (PNES).

Most clinicians have their offices and research labs on the second floor of the building. But there's also one special room that is different from the EEG-monitoring rooms on the lower floors: a hybrid studio for making videos, including EEG video monitoring, as well as for conducting seizure provocation tests. Equipped with cameras, lights, and assorted medical paraphernalia, this room is busy with nurses walking in and out to check on a patient and an EEG technician who closely monitors brain waves throughout. But the room also features the ominous and unusual presence of a large mattress on the floor to soften the patient's anticipated collapse. Last but hardly least, a doctor in a white coat holding a syringe is ready to administer a strange-looking pink concoction into a patient's IV line. Historically, doctors used to inject patients with a pink mixture of vitamin B_{12} and other wholesome goodies; nowadays, they add the pink touch as a colorful dye to jazz up the clear saline for dramatic effect. If a seizure occurs, of course, the medical team is ready to immediately respond with the necessary measures.

What can these clinicians and researchers learn about seizures and suggestions from this unusual, literally provocative intervention? First, they can get a better clinical handle on the problem at hand and make a more accurate diagnosis. Epileptic and psychogenic seizures call for different treatments; the former respond well to antiepileptic drugs whereas the latter don't. And yet, unfortunately, it's a sad fact that many individuals with PNES unnecessarily receive antiepileptic drugs, sometimes over many years—even decades.

Rarely would persons diagnosed with PNES benefit from anti-seizure medication; if anything, such drugs can sometimes worsen their condition. Moreover, no medication has been proven to treat PNES. The confusing part is that many of these patients present a positive response to antiepileptic drugs, either because these individuals are suggestible and expect to get better or due to improvement in the psychiatric comorbidity—say, depression—that often accompanies their condition. Such a positive response further contributes to masking the correct diagnosis (of PNES) and delays providing these individuals with effective treatment options.

This provocation test better illuminates whether a patient suffers from epileptic, psychogenic, or some mixture of these different seizures. An accurate diagnosis can spare individuals with psychogenic seizures time, expensive and ineffective treatments, and unnecessary tests that are less likely to be meaningful for their specific condition. Because abnormal brain electrical activity isn't the cause, effective treatment of PNES typically addresses the underlying psychological problem or psychiatric disorder. It usually involves a multidisciplinary team that includes neurologists, psychologists, psychiatrists, nurse specialists, and primary care doctors. Together with other mental and behavioral health providers, such as social workers and licensed mental health counselors, these clinicians can deliver psychological therapies from cognitive behavioral therapy to hypnosis and even a form of neuropsychoanalysis (an integration of neuroscience and psychoanalysis, a psychological framework that aims to treat mental conditions by investigating the interaction of conscious and unconscious elements in the mind). It seems fitting that this collective clinical effort is taking place on a street named after Sigmund Freud, a neurologist, founder of psychoanalysis, and master of suggestion.

Suggestible Asthma?

In the late 1800s, Dr. J. Noland Mackenzie used a form of clinical deception to test whether one of his patients suffered from asthma attacks—coughing, wheezing, and troubled breathing—due to biological factors or circumstances that were merely psychological. Mrs. X showed asthmatic symptoms and had reportedly experienced sudden attacks as a function of exposure to pollen. But Dr. Mackenzie suspected something was amiss and decided to present Mrs. X with an artificial rose masquerading as a genuine one.

When she entered his consultation room, she expressed that she was feeling well without any special symptoms. But when Dr. Mackenzie displayed the rose, Mrs. X immediately started to feel unwell. She began to complain of tickling in her throat, sneezing, obstruction of her nasal passages, and even congestion in her chest. Upon the revelation that the rose was artificial, Mrs. X was aghast; she left feeling hopeful that Dr. Mackenzie could cure her asthma.[12]

Mrs. X probably displayed a case of hay fever—an allergic reaction to airborne substances, such as pollen.[13] But her story illustrates how asthma, which we often consider through a biological lens having little to do with psychology, may be susceptible to mindset and expectations. In 1885, Dr. Mackenzie intimated to Mrs. X that by smelling his mock flower, she may develop an asthma attack. Little did he know that his nineteenth-century rose ruse would herald the provocation tests that some asthma centers worldwide put their patients through today.

When I studied lung physiology in the late 1980s, my teachers taught me that asthma is a biological condition. Nowadays, in coming up with an accurate asthma diagnosis, some experts recommend checking the lung responsiveness to airway contaminants.[14] In other words, they recommend that patients come in when they're feeling fine—not when they're having an asthma attack—and go through a panel of increasing exposures to an airway irritant until an asthma attack surfaces. This gradual introduction aids in assessing how "twitchy" the airways are by examining how much irritant it takes to trigger an asthma attack.

This method may sound crude, but it's actually a very effective way of ascertaining the resilience and robustness of the lungs and airways to irritation and exercise. A trained respiratory therapist carefully monitors the procedure, and the results help pulmonologists to plan the most effective treatment. Once the doctors notice the incipient signs of an asthma attack, they immediately stop the challenge, provide the necessary treatment, and relieve the symptoms by restoring breathing to a normal baseline.

In this kind of bronchial challenge, patients usually inhale increasing doses of methacholine/histamine—a drug that causes airway inflammation and narrowing comparable to what we see in asthma attacks—or mannitol, a drug that dries up the airways. Typically, the respiratory therapist will prepare five increasing doses of the irritant. Following each administered drug dose, patients perform a breathing test to assess whether and to what extent their airways constrict. My students and I had an idea about these challenge tests. They reminded us of psychogenic seizures and the rose test by Dr. Mackenzie. So we came up with a research idea.

I asked my colleagues at the Montreal Children's Hospital how many kids got an asthma attack within the administration of either the first or second dose of the irritant, and I was glad to learn that most patients showed asthma symptoms within the first two doses.

"Most times I don't even get to the third—let alone fourth or fifth—dose because the vast majority of children develop a full-blown asthma attack following the first or second rounds," one of the nice respiratory therapists educated me. "Over the past thirty years, I have had to administer a third dose only once."

"If that's the case, why don't we make the first two doses into sham placebo dummies and make doses three, four, and five the irritant equivalent of doses one, two, and three?" I asked a senior pediatric pulmonologist, who immediately understood where I was going with this question.

"If we do this," I continued, "I bet you we'll still get most kids to develop an asthma attack during the first two placebos."

He hesitantly nodded in slow agreement.

"And if we don't"—I winked at the pulmonologist—"you'll still have the third, fourth, and fifth doses to test the effects of the airway irritant as in the original first-three-dose challenges."

Thinking You Ingested an Airway Irritant

The attending pulmonologist gave me his blessing, as did the IRB—that watchdog committee designated to see that research proposals using human participants obey federal regulations—and our pilot study was soon underway. In line with our hunch, for the children who came in for our modified bronchoprovocation test, the conceptual expectation associated with ingesting an irritant got most of them to report an asthma attack during the first two (placebo) doses.[15] Objectively, the measurements by the respiratory therapists showed their airways remained largely intact,[16] which meant that the children's feelings were entirely subjective.

On the one hand, these children were under the impression that they were having an asthma attack. On the other hand, the medical instruments objectively showed that their lung function was fine. In other words, they manifested asthmatic behavior and symptoms— similar to those observed when the airways get inflamed and shortness of breath occurs—although their airways remained pristine! There was never any irritant in their lungs.

Although, due to ethics, I never studied the reverse condition— in other words, what happens when patients receive an airway irritant with the suggestion that they are getting a placebo—I decided to further elucidate the discrepancy between objective and subjective observations by running additional experiments. This time I decided to focus on another biological condition, which involved a sizeable psychological component.

Ingesting Lactose but Thinking You Didn't

Lactose intolerance refers to the inability to digest the milk sugar known as lactose, a component of milk and dairy products. A global phenomenon, about 65 percent of all people cannot efficiently digest lactose after infancy.[17] When an individual manifests a lactose intolerant episode—whether an allergic reaction or sensitivity to lactose— symptoms may include diarrhea, bloating, flatulence, and abdominal pain. Some ethnic groups show more symptoms than others. For example, Asian and Jewish people have decreased amounts of lactase in adulthood—that makes them more lactose intolerant—compared to their Caucasian counterparts, especially those of northern European descent.[18] In the same vein, Indians who typically consume larger amounts of milk and dairy products are more tolerant of lactose compared to, say, the Chinese.

By her own admission, Jess—one of my Jewish friends—was "seriously lactose intolerant, like the rest of the family." She stayed away from dairy products and visited her GI specialist faithfully. She's had her share of breathing tests after ingesting lactose-rich liquids, along with stool samples, genetic checks, and other evaluations, so her lactose intolerance was indubitable and medically verifiable. One day we were strolling along a promenade when we happened upon a gelato stand.

"Would you have a gelato with me?" Jess asked.

"Sure!" I answered with surprise. "But what about your lactose sensitivity?"

"Gelato is not ice cream," Jess explained. She was technically right: "It's Italian-style ice cream with minimal air incorporated into it, which creates a dense texture of intense flavors." But then she added a surprising ending—"It has no dairy."

"If you say so, it's my treat!" I said, and bought us two healthy portions, which we enjoyed thoroughly.

I am no big maven of frozen desserts; I'd feast on ice cream, sorbet, sherbet, frozen yogurt, and yes, even gelato—all the same. Still, as far as I knew, gelato contained less fat than ice cream because of the type of milk and cream used in its preparation, but I could've sworn it contained dairy—and lactose—because it had milk in it. On the other hand, Jess had her own sources and notions. She was certain that gelato was lactose-free, so, as far as she was concerned, gelato was safe for her to enjoy. I was intrigued by how she exhibited no symptoms at all having devoured a generous portion. After all, no matter how you lick it, that specific gelato contained lactose—I know because I actually went back and confirmed with the gelato man.

In 2013, my McGill graduate students and I conducted a study examining the psychology of lactose intolerance. Together, we predicted that if individuals who identified themselves as lactose intolerant, just like Jess, learned that they had ingested lactose, their symptoms would surface as a function of this knowledge, but not before. We came up with an experimental design that was approved by the IRB and entailed giving participants—who had either self-diagnosed as lactose intolerant or been diagnosed by their physicians—a pill containing the substance β-D-galactopyranosyl-(1→4)-α-D-glucopyranose.

Pharmacies across Canada used to sell this pill as an over-the-counter placebo. Unbeknownst to most, however, this fancy, long, technical term is just the chemical name for lactose! Omitting "lactose" from our narrative, we explained to participants that they were getting a placebo pill. We were giving them lactose but used the chemical name for it and explained that it was a pill available as an over-the-counter placebo—all true but selectively cheeky.

In line with our hypothesis, most of our participants showed no symptoms despite ingesting this placebo lactose pill. However, long after symptoms should have appeared, and immediately after we made participants aware that they had actually ingested lactose, their symptoms surfaced. At least for some people, expectancy and mere thought seemed enough to either inhibit or facilitate symptoms. These participants showed little by way of symptoms as long as they were unaware of the lactose. But as soon as they realized that they had lactose in their system, they broke out with a whole spectrum of clinical signs.

Not Taking Lactose but Thinking You Did

A few months later, in a separate pilot study with different participants, we predicted that if individuals who considered themselves lactose intolerant learned that they had ingested lactose, their symptoms would surface as a result of this knowledge *even if they hadn't actually consumed any lactose.* To achieve this effect, we had to come up with another crafty ploy.

We partnered with a major milk producer in the province of Québec and obtained a bunch of empty, sterile milk cartons, which we filled with lactose-free milk from the same company. We then invited people who self-identified as lactose intolerant to have as little as the thin film that clings to a teaspoon after you empty it, or more, to see how much lactose it takes to bring about their symptoms, like the asthma-challenge procedure we discussed earlier. Here, of course, participants were unaware that the milk was lactose-free; instead, they went by the outside label advertising it as regular (i.e., lactose-containing) milk. As you can imagine, what people *thought* they had ingested appeared more relevant than what they had actually ingested. Although the milk was always lactose-free, the larger the amount of milk they had agreed to drink, the more dramatic were their symptoms.[19]

So, is lactose intolerance biological or psychological? Clearly, biology plays a major role in epilepsy, asthma, and lactose intolerance. But it seems that knowledge from the medical and life sciences alone isn't enough to explain these processes all the way; we must consult psychology and social science for their important contributions as well.[20] The take-home message should be clear: what we think (top-down influence) seems to interface with the bottom-up influence that we often construe as purely biological.

Deception Impacts Physiology

Anne was a discerning, athletic woman in her late thirties or early forties. Her blue eyes and recessed dimples gave her a distinct touch of finery infused with European aristocracy. The psychiatry clinic was busy that day, but even with a quick peek through the one-way observation window at the reception post, I could tell that Anne was unlike the other people in the waiting area. Wearing her sunglasses high on her head, she was systematically leafing through a stack of magazines that she had culled from the other tables. But she wasn't reading—she was quickly paging through them, pausing to look for long seconds at each image that depicted children and babies. She seemed to get a frisson of excitement from every such photo.

With this observation in mind, I scanned Anne more closely. I could glimpse in her pocketbook the top corner of the ever-so-popular *Baby and Child Care* by Dr. Benjamin Spock. A quick last glance at her torso confirmed a protruding belly on top of her lean frame. When I called out her name and invited her in, she closed the journal she was looking through, got up heavily with her hands strangely protecting her abdomen, and walked toward me with an introductory smile.

"Good morning," I greeted Anne joyfully. "What brings you here today?" I asked, although the referring physician clearly marked the issue in her chart.

Anne spoke openly about her problem. She'd had a yearning desire to be a mother from her teenage years and throughout her fifteen-year-long marriage, but only recently was she able to finally get pregnant.

"That sounds like good news," I said. "What's the problem?"

"Well," Anne mumbled, trying to find the best way to describe it, "I'm already in my second trimester…"

"How are you feeling?" I asked, making room for her to expound.

"The morning sickness of the first trimester is already a bit better. I don't spend my mornings hugging the toilet." She smiled with relief.

"That's great!" I led her on. "What else?"

"Well, I hope it's a girl," she said.

"What does your doctor say?" I probed further. "They can probably run some tests. Did you ask your OBGYN?"

"I don't really want to find out," she replied, "I prefer it to be a surprise."

"I see," I said softly. "When was your last ultrasound?"

The pause that followed signaled that we were getting closer to the critical point. "I don't really need all these gynecological procedures," Anne finally blurted out. "I'd like things to unfold naturally..."

"Unfold naturally how?"

She finally spoke her mind. "Just proceed with a regular, normal, healthy pregnancy without complications and unnecessary procedures and tests."

It took a bit longer until Anne felt comfortable to share with me her secret: "As you can see for yourself, I am clearly pregnant," she said with her hands around her bulging belly. "Things are going well, and I'm not going to let oblivious gynecologists mess me up just because all those faulty pregnancy tests keep on showing negative results."[21]

* * *

It wasn't the first time that I'd heard of such strange behavior, but it was the first time that I had encountered it in person. Another patient I'd seen with a similar, but different, condition was a teenage girl who was certain she had cancer that no doctor or clinical test was picking up. She kept going from one oncologist to another, a string of endless appointments, complaining about symptoms, amassing lab orders, and booking medical procedures. That case was unusual, but Anne was equally special.

Pseudocyesis refers to a false, imaginary, hysterical, or spurious pregnancy.[22] It's a rare condition in which an individual has all the signs of pregnancy, such as abdominal distension, breast enlargement, pigmentation, morning sickness, and even, in some cases, an elevated level of pregnancy hormones—all except the confirmation of a fetus.[23] False pregnancies have deceived women, families, and doctors for decades.[24] Even more mind-blowing, it can also occur in males. In the odd case that it does, major psychiatric problems are typically present.[25] As you can imagine, pseudocyesis correlates with mental disorders,[26] and treatment requires the cooperation of gynecologists, psychiatrists, and psychologists.[27] Beyond humans, moreover, researchers observe false pregnancies in nonhuman animals—for example, mice, rats, wolves, and monkeys.[28] The evolutionary aspect of such behavior may thus be more meaningful than simply dismissing it as abnormal psychiatric phenomena.[29]

Whereas abdominal enlargement often relates to weight gain and gas accumulation,[30] breast enlargement, pigmentation inconsistencies, and even a positive pregnancy test typically relate to the endocrine system, which is susceptible to psychological parameters.[31] Thus, clinicians consider false pregnancy as a psychosomatic disorder provoked by irregularities in the endocrine system.[32] In pseudocyesis, the distended abdomen may remain bloated for months and only vanishes after the individuals recognize their state or when under anesthesia.[33]

The "typical profile" of someone experiencing false pregnancy: a childless woman approaching menopause or a woman who is newly married.[34] These women—and in some peculiar cases, men—often suffer from depression.[35] Symptoms of depression, in turn, can result in weight gain, an unhealthy diet, and the ingestion of psychiatric medications, which may lead some individuals to believe that they are pregnant.[36] Alternatively, pseudocyesis can also occur in sexually active young women afraid of becoming pregnant, and later incorrectly thinking that they have been impregnated.[37]

My journey with Anne spanned several weeks of emotional hypnosis sessions to explore and help her accept her nonpregnant state. In a series of forty-five-minute meetings, we investigated different reasons that led her to perceive herself as pregnant. She was highly hypnotizable, and the sessions allowed her to experience a calm, slightly detached, and less-inhibited plane of consciousness. Cumulatively, over these contemplative meetings, she had the time to consider, process, and accept that she wasn't actually pregnant. It dawned on her gradually over multiple sessions spaced over a few months. About two years after Anne had left the clinic realizing that she wasn't pregnant, she came to visit and introduced me to her cute, newly born twins. Her real babies provided a nice tangible closure to a rather surreal and imaginary situation.

* * *

What do false pregnancies, lactose-filled gelatos, bronchial challenges, and psychogenic seizures all have in common? In all of these cases, "higher-level" expectations and suggestions somehow influence "lower-level" physiology in some top-down fashion. When I first discovered the field of top-down control and "suggestible physiology," I couldn't help but ask myself, *But how? What's the mechanism? How does it work?* It turns out these phenomena are explainable not just through the lens of deception but through the lenses of dissociation and absorption—two constructs that hold a prominent place in psychological science and that deception can illuminate in interesting ways.[38]

Dissociation refers to an altered behavioral state wherein the integrity of our experience seems removed or detached.[39] Absorption refers to the process whereby a person becomes deeply immersed in their mental imagery.[40] Dissociation and absorption form overlapping sets that shape the basis of how we think about special mind-body interactions—for example, hypnosis and meditation[41] or thought-insertion experiences.[42] While neuroscientists show how the human brain can give rise to dissociation through complex brain wave oscillations,[43] behavioral scientists inform our understanding of absorption through psychological studies.[44]

Studying the crosstalk between dissociation and absorption is on the rise.[45] Humans can naturally and spontaneously dissociate—through daydreaming, for example—without any special training. But sometimes we do so as a result of trauma,[46] epilepsy, or the ingestion of psychedelic substances.[47] Despite the substantial importance to both basic and clinical science, we currently possess only an elementary understanding of dissociation and absorption, with thin leads identifying their underlying mechanisms.[48]

The stories in this chapter might seem to reflect a strange or extreme collection that would occur only in remote psychiatry wards, but many of these situations happen to regular folks. At the end of the day, it turns out that the mental "broth" comprising ingredients like expectation, motivation, attention, concentration, dissociation, and absorption is almost as powerful as my mother-in-law's matchless chicken soup—it plays a key role in propelling our suggestible physiology.

CHAPTER 6

The Reliable Science of Unreliable, Suggestible Memories

B ack in my youth, it was common for kids to stop by the small grocery store in our neighborhood to buy something on their way home from school: some sweets, a popsicle, a donut, or even some milk or a loaf of bread that an elderly neighbor had asked that they pick up for them. You didn't need money; the grocer would write down what you owed on a piece of paper, and at the end of the week/month, a family member would swing by to settle the bill. In those days, our neighborly grocer knew us, our families, and where we lived.

One day, on our way back from school, my mischievous buddy, Urich, and I ended up in our bodega looking around for some treats. The store was busy, and I could see that Urich took advantage of the pandemonium to stuff his coat pockets with some fresh rugelach—crescent-shaped, bite-size cookies made by rolling a triangle of dough around a filling of chocolate—from a sizeable warm tray at the small pastry section. He picked out a few more small items and went to the counter. I, too, got something and threw two pieces of rugelach into a paper bag. As we stood shoulder to shoulder at the register, Eli, our friendly grocer, asked us about school and what we'd been up to. To my right, I could sense that Urich was nervous; he was sweating bullets and stuttered out a jumpy "Nothing much." I felt bad for both Urich and the nice grocer.

"I am working on a new magic trick," I blurted toward Eli, who was about to add up the "unhidden" items that Urich put on the counter.

"And what trick might that be, Amir?" Eli finally asked.

"Check it out!" I announced in a loud voice that got the attention of some shoppers.

I reached into my paper bag and took out one rugelach, showing it around. I quickly put the entire piece in my mouth and swallowed it whole, in one gulp, without chewing. I could feel my eyes tearing up from the instant pressure in my widening esophagus. It felt strange to waste a piece of rugelach like that—without enjoying the chocolaty taste—but before anyone could squeeze in a word, I had already reached for the second piece and repeated the same gobbling maneuver. Eli's pupils dilated, and I immediately showed the paper bag and my wide-open mouth: empty and rugelach-free. Eli stopped punching numbers on his calculator and displayed a confused expression. "Your trick is swallowing rugelach without chewing?" he asked with disappointment. "No, Eli, it's sleight of hand—the hand is quicker than the eye!" I said and theatrically reached into Urich's left coat pocket. I fished out two pieces of rugelach and then triumphantly dropped them, one at a time, into my empty paper bag. The whole maneuver made Eli think that I was able to vanish the rugelach from my paper bag and then retrieve them from Urich's allegedly empty coat pocket after I had somehow moved them in there.

Urich was shocked stiff; for him, this unexpected display of impromptu magic was too close for comfort. I could sense that he was itching to bolt out the door. Eli, on the other hand, was still processing what he just saw. Beginning to end, my strange miming act took only seconds, but from his prime viewing angle above the counter, there was no way Eli could have missed any sleight of hand or suspicious move. He looked at us with smiley eyes and said, "Now how did you manage it so quickly? Well done! Today the rugelach is on me, guys." Relieved, Urich and I quickly left his bodega before he could change his mind.

Even before we had gone through half the generous batch that Urich had emptied from his pockets, my taste was cloyed with the sweet richness of chocolate rugelach. Fueled by the sugar, my mind was racing. I was already on my way to realizing that a good performance—whether in the context of a show, hypnosis session, social influence, transmission of incorrect information, or magical illusion—can render people, even if temporarily, more suggestible. It was the first time I had witnessed firsthand how a spur-of-the-moment trick could communicate a very strong message to another person. Eli "fell" for my visual suggestion, and it saved all of us from a very uncomfortable situation.

I was eleven years old, and Eli was probably in his fifties, but from that day on, he'd always comment on the agility of my fingers. For a while, I did wonder if he was just playing along with a child out of adult responsibility, but then he seemed to propose that he had actually noticed how I only *pretended* to swallow the rugelach and that at least one piece "flashed" when I *transferred* it into Urich's left coat pocket. With human memory being so reliably unreliable, we all suggest ourselves into alternative realities. That's the power of narrative.

Any magician worth their salt knows how to leverage this feature of human memory to turn a mere magic trick into an absolute miracle. People don't remember well, and when they attempt to recollect, they often incorrectly recall, add, or subtract (sometimes critical) information that magicians can strategically leverage to enhance the desired effect. We all have known this our entire lives: human memory is malleable. Sometimes we just cannot trust memories, even our own. It's common knowledge now that suggestion can seed a false memory; this well-documented phenomenon goes beyond magic and entertainment. It entails both psychological and legal implications.[1]

Suggestible Memories

Elizabeth Loftus, my friend and colleague from the University of California in Irvine, experienced the power of suggestion on memory at an early age.[2] When she was fourteen years old, Beth lost her mother in a tragic drowning accident. Although she originally had little memory of the event, one of her relatives told her and insisted that she was the first person to find the body. Coming from an older relative, this suggestion was convincing enough that Beth began to "remember" and embellish little details of the accident—for example, the firemen on the scene giving her oxygen—to the point that she actually believed and came to accept the narrative of her relative. But one day, the relative called to apologize for their mistake. Beth had never found the body—but because of the suggestion from the relative, she made up her memory of doing so.

Today, Beth is one of the most prominent memory researchers. She has spent the better part of her professional career researching the effects of suggestion on memory and has been involved in many cases showing what happens in the courtroom when our memory of events may be a result of suggestion. Beth has provided compelling evidence of wrongful convictions resulting from false memories—for example, memories implanted in the mind of a witness by family members, officers of the court, or persons of authority.[3] She and others show that our memories are (re)constructive, a little bit like a Wikipedia page: sure, you can change it; but so can others.[4]

Suggestive influence can potentially tamper with the memory of a witness. Police officers usually investigate and interview witnesses before any lawyers or prosecutors enter the scene. But many things that police officers do, sometimes inadvertently, can meddle with the original memory of the eyewitness. Consider, for example, an officer who shows the witness six photographs—a six-pack—asking that they identify the "bad guy." The witness looks over the photos and cannot make up their mind; they don't readily recognize the bad guy. So far, so good.

Now let's assume the officer says, "Wait a moment, I see your eyes spending more time on picture number three. Let's talk about that." Well, soon this witness may be thinking, "Okay, number three." Perhaps this dynamic may take place innocently, not even on purpose, but the officer is, perhaps unintentionally, leading the witness. Moreover, sometimes the officer knows who in the six-pack (or in the lineup) is the suspect and can sometimes influence the witness in subtle ways. For example, once the witness makes an identification, even if they're not all that sure, they may say, "I kinda think it's number three." What's the officer to say?

Some officers give feedback: "Good job." "That's our suspect." "That's the one that somebody else picked." "Yeah, that's right!" or "That's the one we have a fingerprint match on." But when you give that kind of feedback to a witness, it bolsters their confidence. Now they're saying, "Oh, yeah, I know it was number three for sure." But that creates a problem: the witness will now be more impressive at trial, more impervious to cross-examination, more influential, and more likely to help get a conviction, *whether the defendant is guilty or innocent.*

Today, most psychology students know that leading questions can contaminate memory. When you show people simulated crimes or dramatizations of accidents and ask them what they remember, a lot depends on how, not just what, you ask. For example, "How fast were the cars going when they *smashed into* each other?" will produce answers that indicate the cars were going faster, and is more likely to elicit recollections of broken glass, compared with the same sentence using the verb *hit*.[5] Moreover, the same kind of interrogation could turn a stop sign into a yield sign,[6] and the bad guy into wearing a brown, instead of a green, jacket.[7] Such lab experiments show how easily distortable our memories are and how tenuous our details of events.[8] Sometimes, just allowing a person to overhear some (mis)information is enough,[9] or having a biased investigator who asks leading questions or mentions erroneous details.[10] I can alter your memory using the misinformation effect,[11] but how far can I go? Could I plant an entire memory into your mind of something that didn't happen? You bet!

Experiments show that we can implant an unauthentic memory in people by telling them a plausible story following an innocent procedure. For example, an experimenter would say something like, "I had a chance to talk with your mother. She told me things about you from back when you were five or six years old, and we want to see what you remember and how it compares to what your mother told us." Then, the experimenter would relate a few true stories, just to create some credibility bonding, only to segue into a false story: "You got lost in a shopping mall; an elderly person found you—scared and crying—and reunited you with your family." Through several suggestive interviews, about 25 percent of participants were agreeable to this suggestion and created a false memory of this nonexistent event.[12]

Sometimes, people even report strange and sensational memories. For example, a woman can complain—in passionate, emotional, and compelling form—that her husband brutally cut her baby from her belly. However, if neither observable scars nor any other kind of physical evidence exists to support her story, we will likely flag her memory as questionable. What do you do with memories of horrific experiences, painful tortures, and bizarre abuse when you cannot corroborate the information? Where do such memories come from?

The sources abound: delusions, psychotherapy, imagination exercises, dream interpretation, hypnosis, exposure to false information, cannabis,[13] or simply magical illusions can all lead people to develop elaborate false memories.[14] These endless possibilities of suggested memories beg the next question: Are some people more resistant to or less likely to fall for these "manipulations"?

Researchers looked into this question by, for example, examining people who show the unusual ability of extraordinary personal memory. These highly superior autobiographical memorists (HSAMs) can remember just about everything they did every day of their adult lives.[15] They differ from other superior memorists—for example, those who can recite an inordinate number of the digits of π[16]—in that they don't practice mnemonics to remember vast amounts of information.[17] You may think that because of their astonishing ability, HSAMs would be more resistant to false memories; however, it turns out they are as susceptible as controls. In other words, compared to age- and gender-matched participants, HSAMs were just as susceptible to suggestive memory influence as normative (non-HSAM) folks.[18] So, the mechanisms responsible for producing memory distortions seem fundamental and widespread in humans; regardless of how good (or bad) your memory, it's unlikely that you're immune.

Most people cherish their memories and uphold them as representations of their identity; their memories define who they are and where they come from. But we rarely appreciate the astonishing amount of fiction embedded within the corpus of our own memory lanes. We simply can't reliably distinguish true memories from constructed ones.[19] Perhaps you may think that people would be more emotional about true memories than about false ones. But people can be just as emotional about their false memories as their true memories.[20] Even when we put people in fancy brain scanners, we see similar neural signals for true and false memories.[21] Whether somebody is saying something they deem as true or something that is actually true, we still cannot reliably classify the memory as true or false,[22] at least not yet.[23]

But a century of research about the interaction of suggestion and human memory has taught us one important thing: memory is malleable. It involves reconstruction and occasionally is susceptible to even complete fabrication, and false memories are ubiquitous, sometimes occurring spontaneously and sometimes as a result of external suggestion. That's why we have to be so careful: just because somebody tells you something and they say it with confidence—conviction, eloquence, lots of detail, expressive emotion, and the optics of veracity—you still cannot be sure whether it really happened. Instead, you must resort to independent verification. Perhaps this realization should make you more accepting when you encounter common memory blunders among your friends and family. One more thing: you can plant entirely false memories into the minds of otherwise healthy people—*they need not be highly suggestible*. As Beth Loftus says, "Memory, like liberty, is a fragile thing."[24]

From Hypothetical Questions to Powerful Suggestions and Established Memories

As a teenage magician, I came to appreciate something extraordinary: the phrasing of a question can influence our underlying psychology, and, subsequently, our answer.[25] For example, if you were a South Carolina voter in the 2000 Republican primaries (John McCain versus George W. Bush), you may have received phone calls that asked "If you learned that John McCain had fathered an illegitimate black child, would you be more or less likely to vote for him?"[26] During jury selection, you may hear, "If you were a member of the jury and the defendant was eligible for a death sentence, would you be willing to convict the defendant?"[27] These hypothetical questions are extremely problematic, but pollsters use them knowing they can change people's attitudes and perhaps influence voting behavior. In the first case, given the long history of racism in the South, the push poll assumed that the person answering the phone had racist tendencies, or strong views on having a child with someone other than one's spouse, whether they were willing to admit that over the phone or not. The poll's purpose, however, wasn't to get an honest answer out of the voter; it was to plant an idea in their mind that McCain had an illegitimate love child, even though it wasn't true. In the case of jury selection, by mentioning the death penalty, the attorney is leading the potential jurist into thinking that the death penalty—or guilt—is part of this trial. As abhorrent as these examples may be, they are both examples of push poll techniques that have happened time and time again. Here's how it works:

Usually in a survey or poll situation, you *give* the information—about your opinions, who you are likely to vote for, how many sexual partners you have had, etc. One sneaky kind of survey, however, reverses the situation: A "push" poll *gives you* information, typically with an ulterior motive. Moreover, it typically feeds you some distortion of actual facts.[28] The idea is to push respondents away from the opponent and pull them toward the sponsor of the poll by using hypothetical questions that manipulate facts and sway opinions. For example, in the case of the South Carolina phone calls, John and Cindy McCain *legitimately* adopted their youngest child, Bridget, from Bangladesh. But this example illustrates how the posed question purposefully and malignantly suggested a false narrative about the late senator, which could substantially alter public opinion—especially among people sensitive to such innuendo and racial undertones. To be sure, this misrepresentation was an especially effective slur in a Deep South state where race was still central. While I neither condone nor endorse such nauseating push poll tactics, it's important to recognize their destructive efficacy; McCain lost the South Carolina primary on February 19, 2000, in one of the nastiest, most brutal battles of US politics.

Why do political campaigns resort to this type of smear tactic? Because humans are suggestible...and push polls work! But that was back in 2000, when our sense of truth was different. Fast-forward twenty years, and we don't even need the question to be hypothetical: "If you learned that Donald Trump had misrepresented his net worth for tax purposes, lied about the Georgia election, and sexually assaulted a woman, would you be more or less likely to vote for him?" is unlikely to come close to the original McCain question. Yet, under certain conditions, such questions can effectively influence our attitudes, and subsequently, our behaviors and memories.[29] You better believe it: regardless of your intellect and pedigree, push polls can impact your "truth" and shape your thoughts, memories, and actions. I remember a conversation I once had with a person from South Carolina who vehemently argued that McCain fathered an illegitimate child. It took place after the senator had already passed away, more than eighteen years following the original push poll. These deceptive methods are unfamiliar to most people, and if you don't know about them, you cannot protect yourself against them!

*   *   *

Once, before one of my nightly performances, I attended a communal brunch hosted by one of my friends. During the noisy gathering, my host friend told me that he wanted to introduce me to one of his out-of-town guests. As we stepped away from the main group huddled around the smorgasbord, I couldn't help but marvel at the sight of his red-headed friend. His hair had such a distinct, flame-like hue, it would have made Prince Harry blush.

"This is my buddy, Amir. He can read minds!" my host friend quipped by way of introducing me.

"Can you really?" asked Shir-Li (pronounced *shER-lee*, just as in the name Shirley). His impressive freckles and general demeanor signaled to me that I might have found a suggestible participant for my night show.

"Have we met before, Shir-Li?" I asked with theatrical inflection.

"No. I don't think so," he answered genuinely.

"Well, although we haven't met before today, do you believe in *déjà vu*?" I asked.

"Sure," Shir-Li answered with a smile.

"Great, for today, *déjà vu* is all you need."

I ignored the perplexed expression on his face and moved right along. "Shir-Li, I'd like to try something special with you. Are you ready?" I gestured rhetorically.

"I am!" Shir-Li answered with keen anticipation.

"Now, Shir-Li, if you had to freely choose a figure, any figure you desire, what would it be? Picture any shape in your mind." Before Shir-Li had a chance to answer, I continued, "Obviously, you can choose any preexisting, recognizable shape from any set of figures, or you could also come up with your own new, never-before-drawn figure. It's really up to you."

Shir-Li nodded, trying to decide on his figure. His eyes were darting around as he was mentally scanning the possibilities.

"You basically have an infinite domain of possibilities from which to choose or create your very own, freely chosen personal shape. It should be your free choice." I paused.

Shir-Li nodded again, and I could sense he was getting ready to make up his mind.

"I'd like to show you something interesting I can do with your shape, but before I do, I have to confess that given my limited artistic talent, even drawing a heart and a star would be difficult for me. To be completely honest, I think I have a real spatial deficit with sketching—some sort of 'graphical dyslexia.' Even simple geometric shapes require me to draw and redraw several times. I can do a decent square and even a rectangle, but please don't think of any of those. I want the shape to come from you, not from me. Just think of *two* simple geometric shapes, okay?"

I paused, letting all this verbiage sink in. Shir-Li closed his eyes and concentrated on the two shapes in his mind. "Please keep those thoughts in your head for me. Etch them deeply into your memory, and don't tell me or share with anyone else. I need to make sure you have them locked in so that you can recall them later with ease, and—

"Hello! Long time no see. How have *you* been?" I suddenly pretended to greet another friend who had just walked in behind Shir-Li's back. Disoriented, Shir-Li abruptly opened his eyes and froze in confusion. "Hold that thought, Shir-Li." I asked him to pardon my unexpected departure. "Forgive me, I will get back to you," I uttered as I disappeared into the brunch crowd.

I was careful to avoid Shir-Li for the remainder of the brunch. Just before I left, I came over to him and apologized. "I am sorry we didn't get to complete the demo. May I make it up to you with complimentary tickets to my show tonight?" I handed him two VIP comp passes to the evening show. "I think you'll find it interesting."

"Thanks," Shir-Li said looking at the tickets I placed in his palm, "I still have to figure out my evening plans..."

That evening, his striking hair stood out like a bright torch amid the dimly lit crowd. When I invited him to join me on stage, casually and apparently by chance, I made sure to engage in cheesy small talk that everyone could hear:

"Remind me of your name?" I inquired as I would of any random person I invited onstage, adding, "I'm the rhyming mind-reader, and *surely* it's not too *early* to get to know…"

"Shir-Li," he replied in bashful rhyme, his face turning a shade closer to that of his hair.

"Of course, *Shir-Li*," I said, which got a few laughs from the audience.

"We have never met before today, have we?" I asked, urging him to affirm to everyone in attendance what was technically correct.

He verified my statement for all to note and came onstage.

"Well, if we haven't met before today, do you believe in *déjà vu*, Shir-Li?" I segued into familiar territory: "Now Shir-Li, if you had to freely choose a figure, any figure you desire, what would it be? Picture any shape in your mind. Obviously, you can choose any preexisting, recognizable shape from any set of figures, or you could also come up with your own new, never-before-drawn figure. It's really up to you."

Shir-Li looked at me knowingly; he recognized this pitter-patter.

"You basically have an infinite domain of possibilities from which to choose or create your very own freely chosen personal shape. Yada yada yada." I smiled at Shir-Li and handed him a sheet of paper on a clipboard.

"Please forgive me, this is terribly embarrassing because I am the rhyming mind-reader, but with all the *hurly-burly* I have forgotten your name…"

"Shir-Li," he said again, this time with mild annoyance.

"Of course, Shir-Li. Do you believe in *déjà vu*?" I waited for the laughs to subside.

"First, before you sketch the *one* shape of your choice, please concentrate on what you are about to draw. Use your mind and focus on the figure with your thoughts. Meanwhile, let me scribble something on my pad here." I went through an elaborate act of receiving his mental vibrations, drawing and erasing with a black Sharpie, then trashing the paper and redrawing, until I finally settled on something. When I was done, I folded the piece of paper and held it in my hand. I then handed Shir-Li the Sharpie. "Now, please draw your freely chosen shape, as clearly and basically as you possibly can. Take your time."

Shir-Li scribbled it down and held the drawing against his chest upon completion.

"Show me and show everyone what you have there." I encouraged Shir-Li to reveal his drawing. He flipped over the clipboard to expose a large triangle.

I slowly unfolded my piece of paper to reveal...a triangle! The two triangles were a bit different but undeniably identifiable exemplars of the isosceles variety. The audience erupted in thunderous applause, and Shir-Li returned to his seat with a look of utter amazement.

*   *   *

Magicians never reveal their secrets; however, to achieve this effect, I had put in some prep work during the morning brunch. In other words, I planted the seeds of the trick long before it was finally performed. The effect draws heavily on a few principles we have covered in this chapter, in addition to some intuitive deception. For example, it should be obvious how I got Shir-Li to acquiesce and verify that we had never met "before" by asking him to confirm that "we have never met before *today*." Obviously, Shir-Li is likely the only person in the room, other than me, to savor this linguistic subtlety; but even he didn't blink. For the audience, our dialogue signaled that we had never met before. Period.

But how did I lead Shir-Li to choose an isosceles triangle? Well, that's a bit more nuanced. First, the instructions I glibly rolled off my tongue during brunch were more detailed and more directive than the open-ended description I used at the show. The two descriptions started off the same way, suggesting to Shir-Li that they were identical, but during the show I stopped short of going into all the constraints that would lead Shir-Li to converge on two simple geometrical figures. In other words, the audience didn't get to hear that part, but Shir-Li "heard" it by implication, or "understood" it by intimation.

I was using the opening bit of this pitter-patter to trigger Shir-Li's memory so that he would recollect the two simple figures he had settled on during brunch. The French expression, *déjà vu*, suddenly becomes more meaningful to him. But I also need to justify *déjà vu* to the audience, so I play ditzy and feign a cognitive lapse. The spectators then attribute *déjà vu* to the repeated comical situation of asking for his name.

Second, what two simple shapes would *you* have chosen after I had previously contaminated *heart, star, square*, and *rectangle*? If you run a small experiment, you will discover that when confronted with this situation, most people come up with either a triangle or a circle. It's just natural to settle on one of these two common geometric shapes—perhaps because they are easy to outline, immediately pop into mind as the simplest, or just because other options are less salient. Remember, I'd made it abundantly clear that I was a hopeless case when it came to drawing. During the show, I also made sure to remind Shir-Li of this fact by sketching and resketching with tremendous effort what should be a very simple figure. The audience, on the other hand, likely attributed these difficulties to my mind-reading abilities; they certainly didn't know that I instructed Shir-Li earlier in the day to choose a *simple geometric shape*. On the contrary, the audience is under the impression it could be *anything* Shir-Li wanted to dream up. Moreover, the passage of time since brunch and the dynamic of the show further obfuscate what Shir-Li may have remembered and later reconstructed about his experience.

Finally, when I first implanted the whole preamble to this trick during brunch, I used a variation of a well-known, conversational mime-gesture technique to influence Shir-Li into choosing a triangle.[30] Of course, all of this hardly guaranteed that Shir-Li would go for a triangle, let alone an isosceles; for now, let's call this last part "magician's luck."

* * *

I was able to reshape Shir-Li's memory of the event by harping on with my continuous commentary and by blurring my suggestive messages to him with those meant for the audience. Walking away from this stage experience, his later recollection was that he had chosen his selection from a virtually infinite domain of shape possibilities, as I suggested to him during the stage performance, and that I never so much as came close to influencing his decision. In his mind, our earlier brunch interaction that day was meaningless, irrelevant, not even worth mentioning. Moreover, when spectators surrounded him after the show to inquire about his experience and perspective, he reaffirmed that he had not met me before that day and that I never directed his selection process. His post-hoc ascriptions made him pass as a dependable, trustworthy source. "It was my free choice," he said with conviction, "I have no idea how he did it. Amazing!"

Not only do our brains not work like video recorders, but suggestions can manipulate, corrupt, impair, alter, or even reconstruct whole-cloth memories. As we have seen, people who witnessed a car accident may swear under oath that a stop sign was a yield sign if suggestion tips them to this idea after the fact. Across the three stages of memory—acquisition, retention, and recall—we are most susceptible to the corrupting effect of "post-event suggestion" during retention. As memories fade and weaken, they become more vulnerable to contamination, especially by someone probing with an agenda, media interviews, and leading questions.

Moreover, not only is human memory malleable, but suggestion can shape our decisions and influence our behaviors—sometimes in the most dramatic ways. With a bit of pre-show work and the right kind of showmanship, intense effects can follow.[31] Suggestion can work wonders not only in intimate, one-on-one situations, but also with large crowds and groups. With the right advance work, a suggestion for a violent assault, encouraged by a sitting president, can steer a marauding mob to storm the Capitol. Believing that they've been deputized by their president to stop a crime—the election certification—and defend their country, a fierce crowd attacked the seat of democracy that their president had vowed to protect![32] We have seen variations on this theme in different parts of the world, throughout history and continuing today. Being aware of push polls or related tactics, and having the awareness that suggestibility plays a role in this mix, goes a long way toward protecting us from their influence. Such vigilance is important not just as a potential solution, but as a practical takeaway from this chapter. When someone suggests something inaccurate or disagreeable to you, call it out loud and clear; acquiescing may result in a gaslighting dynamic and manipulation that will be difficult to undo later.

CHAPTER 7

Do Antidepressants "Work" Through Suggestion?

In the early 2000s, I was a young assistant professor of psychiatry at Columbia University in Manhattan. My office at the New York State Psychiatric Institute—informally referred to as the "PI" (pronounced *pea-eye*)—was right next to one of the main waiting areas in the Division of Child and Adolescent Psychiatry, a nexus for parents and their children, with constant coming and going and plenty of boisterous waiting in between. Faculty around this area worked with their doors closed to block out the ambient commotion and reduce some of the noise pollution—but I didn't. Having grown up around a rambunctious home day care, I didn't really mind the hullabaloo. As a researcher, I had to go in and out of my office frequently to check on experiments and students in other rooms, so I just kept my door wide open throughout most of the day.

From my office, I could see and hear countless parents sitting with their children, waiting long stretches for their appointments with the PI psychiatrists. Often, a bored kid would wander away from the waiting area and curiously peek into my office. Some would try to strike up a conversation, anything from "Can I draw on your whiteboard?"

to "Are you also a child psychiatrist?" If I could spare the time, these casual lines would develop into meaningful exchanges. And so I developed many informal relationships with my young visitors.

When they would finally leave my office to head for their appointments, within a few short minutes they'd usually pass by again, this time waving goodbye on their way out. After an extended wait, their actual appointments were usually brief. It was obvious that some of them suffered from mental health problems and needed help; others, however, felt like "bubbly and effervescent" children to me. Given that on many occasions I'd spend more time with them in my office than the psychiatrists would in theirs, I was surprised to discover that it was nearly impossible to see a child leaving the building without a prescription for psychiatric medication. In line with modern psychiatry, moreover, those prescriptions largely fell into three broad categories: antidepressants, antipsychotics (and mood-stabilizing drugs), and stimulants.

As a neuroscientist, I knew a bit about these drugs, their mechanisms of action, and their side effects: I was well-read on the medical and scientific literature and was in the process of developing a special interest in antidepressants. For example, knowing that most antidepressants were not FDA-approved for children, I asked my psychiatry colleagues some tricky questions: "How do you know how much antidepressant a child should get?" Responses would vary, but the main sentiment, albeit articulated obliquely, intimated that children were sometimes treatable as "mini adults." Too often, dosage followed physical metrics (e.g., an 80-pound child would receive about half the dose of a 160-pound adult) and sometimes age (e.g., a twelve-year-old would get about 50 percent of the dose of a twenty-four-year-old patient). I was struck by this off-label application.

This rationale made little sense to me because the mental,

emotional, and intellectual ability of children isn't half of what it will be when they're twice as old or twice as heavy; these abilities seldom grow linearly with either age or weight. As I further perused the antidepressant literature, I quickly discovered the "antidepressant wars"—a debate about the effectiveness of antidepressant medications—a new term for me at the time but one that's still as relevant today as it was back then. And the wars are still raging on.

*   *   *

I was ill-prepared to discover the decades-long influence and financial grip of industry money on academic psychiatry: the pharmaceutical industry paid researchers at brand-name institutions to talk up drugs at seminars and conferences;[1] Big Pharma paid for "expert panels" to promote the use of their drugs; and more often than most people realized, outside firms had written up studies, manipulating data and spinning out interpretations.

To me, this situation made it difficult to make sense of psychiatric drug studies, but most of the psychiatrists around me viewed things through a different lens. My colleagues in the psychiatry department would speak about all kinds of medications and invite clinical researchers to share their findings at prestigious Grand Rounds—a formal type of lecture featuring newsworthy contributions. After attending these Grand Rounds forums, I gradually developed a knack for separating the honest, rigorous presentations from those that masqueraded as research. These latter, infomercial-type shows were usually conspicuous but not always—and without knowledge of the literature, relevant backstage politics, and secret financial handshakes, it would be difficult to know what to believe. Against this backdrop, I began to research and analyze the available data on antidepressants.

To my surprise, I gradually discovered that standard medical care for depression provided about the same response as a placebo. Psychotherapy works just as well as antidepressants, but a combination of psychotherapy and antidepressants isn't much better.[2] What can you do if you feel depressed but want an intervention other than antidepressant medications? Fortunately, treatments for depression, other than antidepressants, do exist: exercise, acupuncture, omega-3 supplements, tai chi, qigong, and yoga seem to provide benefits comparable to those of antidepressant medication.[3] Patients will benefit from a nice reduction in the symptoms of depression with antidepressants, nearly the same relief they will experience with a placebo; however, don't confuse getting placebos with getting nothing. If depressed people receive nothing and just remain on a waiting list, they don't improve—doing nothing doesn't work for depression.[4]

In late 2005, when I published a paper calling into question the effectiveness of antidepressants in children,[5] the chairperson of my department was quick to call my phone extension and invite me for a little informal chat. "I strongly urge you to stick to your academic expertise in suggestion," he reprimanded me. "Leave psychiatric drugs to psychiatrists..." The volume of his voice and the size of his flared nostrils signaled to me that engaging him in a conversation about academic freedom, medical ethics, and the philosophy of science would be pointless.

Around the same time, the FDA held a series of hearings on whether antidepressant drugs—for example, Paxil, Prozac, and Zoloft—backfired in a small number of patients, causing suicidal thinking and desperate behavior. I remember following the hearings with a mounting feeling of incredulity. Hundreds of vociferous family members who had lost a loved one to suicide or got too close for comfort to losing someone to suicide expressed their consternation. Some of them, it was clear, knew at least as much about antidepressants as did the doctors.

By 2006, the FDA had concluded that a so-called black box warning on antidepressant drug labels was necessary, citing the suicide risk for children, adolescents, and young adults.[6] Many of my fellow psychiatrists felt it was the wrong decision, discouraging the use of valuable psychiatric drugs. I could sense that the tension between the inside and outside of psychiatry departments was becoming more palpable.

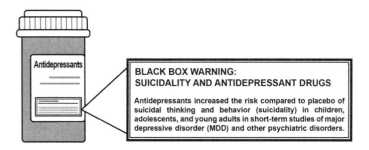

In 1998, Irving Kirsch and his doctoral student Guy Sapirstein published a bombshell study entitled "Listening to Prozac but Hearing Placebo."[7] This meta-analysis (an examination of data across multiple scientific studies) incited the beginning of a tumultuous debate about the efficacy of antidepressant medications. Kirsch and Sapirstein, two psychologists, wanted to examine just how large the placebo effect was for antidepressants. Given that one of the core features of depression is a sense of hopelessness, they theorized that assigning a patient to a new treatment might instill a new sense of hope, thereby undermining this characteristic of depression to some extent. This realization motivated them to analyze data in which individuals with depression received: 1. Active Antidepressants, 2. Inactive Placebos, or 3. No Treatment. The researchers then measured subsequent *responses* and *effects* (see image) from participants in these separate groups.

THE DIFFERENCE BETWEEN
EFFECT AND RESPONSE

Continued on next page →

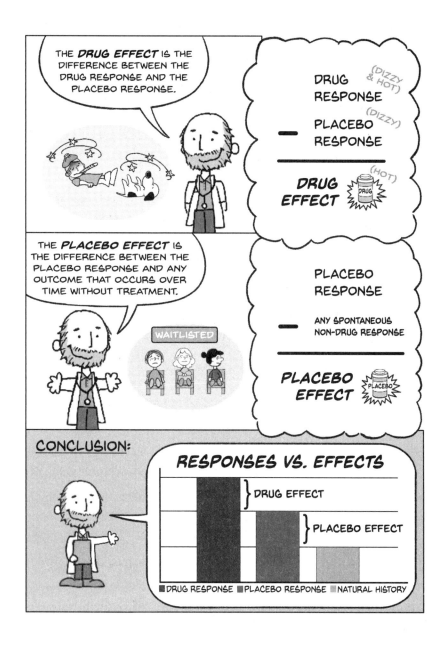

What did they find? Those who received the antidepressant showed improvement: their depression got better. Not a whole lot better, but—on average—reasonably so. Meanwhile, the group that got no treatment at all didn't show much change. That's not so surprising and was in line with what they had expected. But their findings came with a confusing twist: *patients who received inactive placebos improved almost as much as those who got the active antidepressants.*

Given the variety and number of stakeholders, with a special emphasis on Big Pharma involvement and interest, this important study was—and remains to this day—explosively controversial. As expected, its publication met with almost instantaneous criticism. Opponents fired back with several retorts. The most common critical response was that the studies included in the meta-analysis spanned but a minuscule group of unrepresentative, inconsistent, and erroneously selected articles. So Kirsch decided to go to the FDA and collect from them—using the Freedom of Information Act—information about the six most widely prescribed antidepressant drugs from all the respective clinical trials ever submitted in the process of getting FDA approval.

Meta-analyzing FDA data carries special importance in the US because these data form the basis for formal drug approval. If there's anything wrong with these data, then the drug should never have been approved in the first place. To be sure, in line with Section 314.126 in the Code of Federal Regulations, these data must include all "adequate and well-controlled studies."

With their hands on these FDA data, the researchers looked at the response to antidepressants using the most widespread, clinician-administered depression assessment scale, the Hamilton Depression Inventory (HAM-D).[8] With this more complete dataset of both published and unpublished trials, they found that more than 80 percent of the positive responses to the antidepressant were also present among patients who received placebos. Yes, that's statistically significant; however, so small was the difference between the antidepressant and the placebo (less than two points on a fifty-three-point scale) that a question emerged: Is this difference clinically meaningful?

When Do Interventions Make a Real Difference in Our Lives?

Scientists often talk about the difference between "statistical significance" and "clinical significance," but naïve ears may miss this subtlety. Surprisingly, many clinicians and researchers, not just laypeople, also miss this difference.[9] The confusion partly stems from our experience with the evasive nuances associated with the meaning of the word "significance." A common misinterpretation construes the term "statistical significance" as "clinically important" because many people equate "significance" with its literal meaning of "importance." But in statistics and science, this term of art carries a far more restrictive implication.[10]

Back when I was an undergraduate student, some of my professors explained to me that statistical significance is a tool that scientists use to determine whether a specific observation is "for real" or whether it should be dismissible as a fluke, mere coincidence, or an artifact. It took me some time to realize that they were wrong. As I refined my knowledge, I grew to understand that while many researchers speciously construe statistical significance as a statement about the truth of their hypothesis or the probability that random chance produced their data, it's actually neither.

What is it, then? Statistical significance tells us about data in relation to a specific hypothetical explanation but doesn't actually tell us anything about the explanation itself. It's a subtle point but a critical one to savor.[11] Scientists rely on a predetermined, but nonetheless capricious, level of probability and base mechanical decisions around its agreed-upon value of 0.05 to test the strength of evidence.[12] But statistical significance holds

only limited value for sick individuals seeking the best medical intervention, practitioners striving to offer the best clinical treatment, or policy-
makers keen on making the best decision.[13] In other words, it doesn't tell
us much about how practical our findings may be.[14]

On the other hand, clinical significance examines how *big* an effect
is and determines whether the result is substantial enough to make a
practical difference in real life. So, clinical relevance—or importance,
if you will—is an entirely different concept that determines whether
the results of a study are likely to impact our lives. It needs to account
for factors such as how sustainable the effects are, whether clients find
them acceptable, how cost-effective they are, how accessible, and how
user-friendly. In other words, clinical significance also refers to how
easy interventions are to get, apply, or use.[15]

Whereas the scientific community takes great pride in its established
procedures and accepted values to test for statistical significance, it's
largely "out to lunch" when the focus turns to evaluating clinical significance.[16] In my experience, more often than not, subjective judgments by
clinicians and patients—rather than science—decide whether a result is
clinically meaningful. That's not too scientific and probably shouldn't be
the way we come to conclusions, but that's the way things often play out.

The crux is simple: *statistical significance doesn't tell us whether
results are clinically relevant, although in practice clinical relevance is more
important than statistical significance.*

Let's demonstrate the difference between statistical significance and
clinical significance with a thought experiment. A company recruits
one hundred thousand participants to engage in a year-long weight loss
program. At the end of that year, the researchers find that the average
participant lost half a pound. While statistically significant, does losing
half a pound over a year make for weighty news? After all, just how
much weight loss should tilt the scale in favor of palpable health benefits

continues to be a source of uncertainty and debate.[17] Clearly, it would be clinically significant if the average participant lost forty pounds in a year, but would half a pound meet the criteria for "clinical significance"? Studies show that for adults who adopt and sustain physical activity combined with a healthy diet, significant health benefits occur even with very modest or even no weight loss at all.[18] For example, participants following a Diabetes Prevention Program[19] who increased their physical activity for about 150 minutes per week but didn't achieve "clinically significant weight loss" reduced their incidence of diabetes by 44 percent.[20] In this regard, if we apply these findings to the treatment of obesity, it will likely be more clinically significant to focus on cultivating healthy lifestyle behaviors than to focus on statistical significance.[21]

Consider another cheerful example. Imagine a hypothetical Smile Study wherein researchers collect data from a million people to check if smiling improves longevity. After analyzing the data, the scientists discover that, compared to people who don't smile, smiling individuals increase their life expectancy by an average of ten seconds. With such a large sample of people—a million participants—a ten-second difference will likely be statistically significant because statistical significance is sensitive to sample size: the more people you test, the more likely you will stumble upon small differences due to pure chance. But this tiny improvement isn't likely to qualify as *clinically* significant—it's meaningless in real life. No one walks around with a big smile plastered on their face in hopes of extending their life by a mere ten seconds. But ten years—even months—would be a different story.

Statistical Versus Clinical Significance

We often construe statistical and clinical significance as coming together because, in an ideal world, the finding would be effective en masse and valuable to humans; however, that's only one possibility. The weight loss and smile studies show us that it's also possible to find a statistically significant result that lacks clinical significance. In other words, if something is statistically significant, it doesn't necessarily imply that it's automatically clinically significant as well.

In the lucrative world of Big Pharma, for example, drugs that boast both statistical and clinical significance may be a lot more difficult to find than drugs that show just statistically significant effects, especially if you consider an inflated sample size. Given that the lay public isn't conversant with statistics and doesn't typically grasp the difference between statistical and clinical significance, it's easy to overlook the converging findings that researchers and scientists have reported for the past decades. Is it possible that the use of antidepressants for depression carries small statistically significant effects but hardly any clinical significance?[22] One of the most glaring pieces of evidence here is that when the FDA more recently analyzed all the antidepressant trials provided by the pharmaceutical companies, results showed that only about 15 percent of patients benefited from a clinically meaningful response that they would not get on a placebo, and even that might be due to "breaking blind" from the side effects.[23]

In clinical trials and research studies, participants usually don't know whether they receive the drug or a placebo; they are "blind" to the content of the pill. We call it "double-blind" when neither the participant (patient) nor the researcher (doctor) knows whether the pill contains the drug or the placebo. However, if at the end of the study, participants (and researchers) can guess correctly whether they received the drug or the placebo, you know that they have broken the blind. This dynamic usually leads to overestimation in favor of the drug.

With older drugs, for example tricyclic antidepressants, the marked side effects of sedation, drowsiness, dizziness, and dry mouth often contributed to a salient "tell": doctors could easily identify who was ingesting the drug and who was on placebo.[24] During my early days in psychiatry departments, I encountered many patients on antidepressants who had such dry mouths, it would be impossible to miss the conspicuous pattern in which their tongues scraped around and clicked about.[25] Consistently, my colleagues and I were largely able to detect with high accuracy who received the active treatment and who was on placebo.[26] More lately, advanced meta-analysis spanning a whole range of older and newer medications showed that side effects might indeed "unblind" not just doctors but also patients, sometimes subtly, thereby resulting in an overestimation of the efficacy of antidepressants.[27]

But if all that's true, it seems that suggestion and expectation play a large role in antidepressants. A cynical take-home message could be: clinically, taking an antidepressant for depression may be equivalent to taking a Tic Tac, but one that costs you two to fifteen dollars a pop and leaves you with sexual dysfunction and a dry mouth. As with other costly drugs,[28] an "expensive" placebo is more effective than a "cheap" one.[29] Plus, from the perspective of the patient, the noticeable side effects that accompany a pill serve as further "proof that it's working."[30]

When Are Antidepressants Clinically Meaningful?

We often want to know not just whether a specific treatment affects people, but how much it does. For that, we look at *effect size*—the magnitude of the experimental outcome. The larger the effect size, the stronger the relationship between the two variables. For example, we may want to know the effect of an intervention on depression because it will show us if the intervention had a small, medium, or large influence on depression.

One cool feature of effect size is that it permits us to compare and line up outcomes from totally different domains. Doing so can give us an intuitive feel for what it means for antidepressants to have a small (say, 0.3), medium (typically, 0.5), or large (conservatively, 1.2) effect on depression. The values for these effect sizes are arbitrary, of course, but many clinicians and researchers use them, and they allow for helpful comparisons.[31]

For example, if antidepressant medications were therapeutic pills for weight loss, growth hormones for boosting height, or IQ poppers for enhancing intelligence, you'd expect the following outcomes in line with the respective effect sizes:

EFFECT SIZE

	Small 0.3	Medium 0.5	Large 1.2
Weight Loss	8.5 lb	14 lb	32 lb
Height Gain	0.8 in	1.4 in	3.4 in
I.Q. Increase	5 I.Q. Points	7.5 I.Q. Points	18 I.Q. Points

Regardless of initial weight, height, and intelligence, it would be remarkable to lose 32 pounds, grow 3.4 inches taller, or gain 18 IQ points, but such effects certainly aren't typically commensurate with the effect of antidepressants on depression. When we realize that for the effect size to count as clinically significant most clinicians require it to be at least medium—in other words, the equivalent of losing 14 pounds, sprouting 1.4 inches, or achieving a 7.5-point increase in IQ—we must acquiesce that antidepressants, at best, yield small effect sizes, if that.

Another way to visualize this information and get a better feel for the clinical significance of antidepressants comes from the world of pain, wherein researchers and clinicians have developed a consensus on how to determine clinical significance using the Clinical Global Impressions Improvement Scale (CGI-S).[32] Using this agreed-upon tool, we can determine whether an intervention can lessen pain in a real, meaningful way. Here is what happens when we apply this tool to FDA data for the HAM-D based on raw data from forty-three trials with antidepressants spanning 7,131 patients.[33]

This visual comparison shows that "minimal improvement" would require a difference of at least seven points on the scale; "much improved" fourteen points; and "very much improved" even higher than that. But the three-point difference that serves as the criterion for clinically significant improvement mapped exactly onto the average improvement seen in depressed patients whose clinicians had rated them as having "no change." In other words, antidepressants didn't make a clinical difference.

Whether patients are severely depressed,[34] either by the definitions proposed by the American Psychiatric Association[35] or otherwise,[36] these results seem to hold well from the unassuming "mild"[37] to the alarming "*very* severe."[38] Moreover, the teeny-tiny average increase in drug improvement over placebo has been confirmed by a number of independent studies.[39] One particular study, an in-house FDA initiative, looked at all of the antidepressants ever approved up until 2016 by analyzing patient-level data from more than seventy-three thousand individuals.[40] The results echoed what had been reported eight years earlier: so small was the statistical advantage for antidepressants over placebos, it made them less effective than initially promised at treating depression.

But perhaps antidepressants are effective for other conditions? For example, in addition to depression, many physicians prescribe antidepressants to treat anxiety disorders. What evidence do we have that antidepressants work well for the treatment of anxiety? Unfortunately, meta-analyses using the same data, as well as other datasets,[41] revealed that antidepressants carry the same (in)efficacy for anxiety as they do for depression.

And yet all this science did little to shake the beliefs of the antidepressant fans, who insisted that... well, plain and simple—antidepressants work. "Doctors and patients know what works and what does not," wrote a distinguished physician in the UK.[42] Another wrote, "Millions of content patients can't be that wrong."[43] I beg to differ.

Have you ever heard of the *argumentum ad populum* (bandwagon) fallacy? Just because the majority thinks that something is real, doesn't make it so. For example, most humans once thought that Earth is a flat disk. (Today, it's no longer the majority that subscribes to this view; however, it seems that with the advance of social media conspiracy–mongering platforms, we may soon resubscribe to this ancient cosmography in droves.) For centuries, practitioners used to treat millions with remedies such as deer tears, reptile blood, and assorted fresh droppings.[44] Spanning bloodletting and trepanation (drilling holes into human skulls), the history of medicine, even as late as the twentieth century, is replete with unusual recommendations: that asthma patients, especially women and children, smoke cigarettes,[45] or that people drink lithium-laced 7 Up[46] and cocaine-infused Coca-Cola.[47]

Today, we know more about how burning thorn apple leaves releases smoke that is rich with antimuscarinic alkaloids, as well as about the psychopharmacology of lithium and cocaine; however, pseudomedical treatments always "worked" for many individuals, including some very important patients. Consider, for example, the twentieth president of the United States (POTUS), James Garfield, who passed away after his doctor displayed an astonishing determination to feed him through his rectum, or Rosemary Kennedy (sister of the thirty-fifth POTUS, John F. Kennedy), who was the infamous recipient of a prefrontal lobotomy.

The bottom line: everything comes down to the subtle difference between a statistically significant effect and a clinically significant effect. Luckily, we can rely on science, clinical trials, and meta-analyses to illuminate this difference. When we do, we can answer the question with little argument: *the difference between antidepressants and placebos is statistically significant but small—too small to make a clinically meaningful difference.* But if that's the case, is it not possible that these drugs exert most of their antidepressive effects through placebo mechanisms, including the suggestibility profile of the individuals who take them? The answer certainly seems to point in that direction.[48]

Suggestion Pits Placebos Against Medical Intervention

If antidepressants were initially the sole purview of psychiatrists and mental health specialists, today, general practitioners, internists, and family doctors prescribe these drugs liberally, making their use even more ubiquitous. You'd think that with all the data concerning antidepressants we should know better, but life is more complicated than that. For example, consider the case of arthroscopic surgery of the aging knee. Not only has research shown that this procedure holds no benefit beyond placebo, but this surgery actually fares worse than placebo surgery—indistinguishable from the original surgical procedure other than the actual intervention. After two weeks, individuals who went through placebo surgery did better than those who went through real surgery; after one year, folks who received placebo surgery were better off than their real-surgery peers for both walking and climbing, and only two years after surgery there was no difference between the groups.[49]

The *British Medical Journal* (BMJ) looked at what had happened since the initial study reporting on arthroscopic knee surgery for osteoarthritis.[50] They found that, despite a replication of the study showing the same basic results, the rate of doing this arthroscopic surgery had actually *increased* rather than decreased.[51] The *BMJ* accordingly issued a clinical guidance: "We make a strong recommendation against the use of arthroscopy in nearly all patients with degenerative knee disease."[52] But knee arthroscopy remains one of the most common orthopedic procedures in countries with available data, and it is certainly popular with most orthopedic surgeons to this day, including for degenerative knee disease. When my mom was in her late seventies and complained about knee pain due to her osteoarthritis, I had a hunch as to what the orthopedic surgeon she consulted with would recommend. You guessed it—arthroscopic surgery.

For those who may still miss the analogy with antidepressants, it may be helpful to recall the story of mammary ligation for angina— a surgical procedure that ties off or clips the blood vessel that supplies the anterior chest in the event of reduced blood flow to the heart.[53] Initially, between 1939 and 1958, doctors considered this invasive treatment as improving blood flow to the heart via collateral circulation. The apparent success of this operation was all over the news.[54] But then in 1959 and 1960, two randomized controlled trials demonstrated that bilateral internal mammary artery ligation was no better than a sham intervention of placebo surgery in the treatment of angina.[55] The results were towering: real surgery showed 73 percent improvement; sham surgery came in at 83 percent improvement. Although the difference between real and sham surgery was not statistically significant, clearly the placebo surgery was no less effective than the real procedure. After these publications, surgeons rightfully abandoned internal mammary artery ligation. But given the dramatic improvement following both low-risk procedures— ligation and sham surgery—should we perhaps explore how to reap real therapeutic outcomes from placebo and suggestion effects? I certainly think so![56]

The Suggestible Brain Plays with Your Mind

Suggestions can wield some impact on health outcomes. But how much? The actual effect depends on many parameters—for example, the patient and their health condition, the treating practitioner, and the special "human chemistry" of the therapeutic encounter—yet it is undeniable. Whereas suggestion can influence pain thresholds, it's less likely to grow hair on a bald head. But how does suggestion work when it does? We can answer this question both conceptually and technically.

We usually trust that if a doctor prescribes a treatment, it will help. However, as we have already seen, sometimes "effective" drugs actually help fewer people than expected—if by "help" we mean keeping a bad thing from happening to someone who would have otherwise suffered it. Specifically, most people don't fully realize that a lot of interventions benefit only a small number of patients. How many people must get an intervention so that one person will get better? The answer hides within the concept of "Number Needed to Treat" (NNT).

Assigning a specific value for NNT follows a simple procedure.[57] Consider postsurgical compression stockings. People often wear these snug-fitting, elastic socks to prevent the formation of deep-vein thrombosis—serious blood clots.[58] Data show that sans stockings, 27 percent of postsurgical patients develop such clots, but only 13 percent of patients wearing the stockings do. The risk reduction is 14 percent (twenty-seven minus thirteen). Dividing one hundred by fourteen, the rounded answer comes out to seven (because we usually like to work with whole numbers).[59] In other words, seven postsurgical patients need to put on compression stockings so that one of them will avoid developing dangerous blood clots after surgery. The flip side of this insight is that six out of seven individuals, more than 85 percent, will *not* benefit from wearing the stockings postsurgery.[60]

Compression stockings are relatively cheap and mostly minimal risk, so doctors recommend them liberally, even though most patients won't benefit from them. But when a procedure is risky, the NNT helps to weigh the benefits of an intervention for preventing bad outcomes against its associated risks and costs. In general, clinical scientists have arbitrarily decided that NNT < 5 indicates a meaningful health benefit, whereas NNT > 15 signals a small net health benefit, if that.[61] The NNT for antidepressants is closer to 5 than to 15. And yet patients may swear by interventions with NNTs far greater than 15, and physicians may struggle to reap the benefits of interventions with NNTs smaller than 5.

Few patients have ever heard of NNTs, and many physicians widely ignore them.[62] But it should be easy to understand that in the same way that someone can get a stroke despite taking statins religiously, not everyone who's at risk is going to have something bad happen to them. The huge gray zone that stretches between these two extreme outcomes—"getting hit in spite of taking all the right precautions" versus "cruising unscathed even when carrying known risks"—presents fodder for the effects of suggestion. Only with careful scientific experiments can we tease apart how much of the effect genuinely comes from suggestion and what part owes to other factors. Not without its critiques,[63] this research niche forms a thriving field that combines the sensibilities of psychological[64] and brain science[65] with those of clinical medicine.[66]

Humans are sensitive to nuance; for example, when you don't speak of cancer "victims" or people who "fight" a "chronic" disease, you may change more than a narrative—you may well alter outlook and physiology. When we say "fighting," we're already acknowledging the adversary is powerful; "chronic" communicates an air of "uncontrollable" and "indefinite." For some people, these linguistic choices represent strong and meaningful messages.

On the other hand, we have to be careful not to lean too heavily on the "sunshine brigades"[67] that bombard patients with inordinate amounts of "positive thinking."[68] Patients who don't get better under such conditions may feel guilty—as if they, themselves, were somehow to blame, perhaps because they didn't try hard enough.[69] Actually, this mindset may interfere with their healing.[70] Specifically, subtleties of language and motivational parameters may inspire but also disempower the very people a practitioner aims to help.[71] The take-home message: the power of words, suggestions, and expectations shows that mindset holds far greater medical weight than heretofore imagined.[72]

Rather than relying on evidence-based, logical, and rational analyses, most humans follow an intuitive psychological process. For example, when participants received identical statistical information—some in terms of mortality rates and others in terms of survival rates—their treatment of choice showed a marked preference toward survival figures. Moreover, this framing effect was not smaller for experienced physicians or for statistically sophisticated individuals than it was for a group of random clinic patients.[73]

SURVIVAL FRAME	MORTALITY FRAME
Surgery: Of one hundred people having surgery, ninety live through the postoperative period, sixty-eight are alive at the end of the first year, and thirty-four are alive at the end of five years.	**Surgery:** Of one hundred people having surgery, ten die during surgery or the postoperative period, thirty-two die by the end of the first year, and sixty-six die by the end of five years.
Radiation Therapy: Of one hundred people having radiation therapy, all live through the treatment, seventy-seven are alive at the end of one year, and twenty-two are alive at the end of five years.	**Radiation Therapy:** Of one hundred people having radiation therapy, none die during treatment, twenty-three die by the end of one year, and seventy-eight die by the end of five years.

As we can see, presentation matters. Super honest physicians may say to their patients: "Clinical science shows that the antidepressant I just prescribed for you should help with your depression; one of five patients with your condition get better."[74] These doctors could be technically correct and may accordingly feel deliriously ethical.[75] But they'd be disabusing, actually robbing, their patients of a powerful dimension—that of hope—which plays a critical part in depression, wherein hopelessness[76] and unspecific factors[77] play a major role in the outcome.

I'd encourage physicians to limit their initial communication to the opening part of the sentence: "Clinical science shows that the antidepressant I just prescribed for you should help with your depression." In depression, the best available therapy includes hope; however, NNTs don't indicate the likelihood that an individual will benefit from a positive outcome *unrelated to the antidepressant.*

"I wouldn't use NNTs to explain chances and risks to the average depression patient," I mentioned to one of my former students, who is now a practicing psychiatrist.

"Why not?" she expectantly asked.

"Because NNTs are helpful for clinicians and researchers, but less so to patients. We don't want the patient to constantly question whether they will be the lucky one out of five to get better." I let my words sink in a bit before I hit her with my closer: "Hope is the magical ingredient that our suggestible brain so desperately desires. And suggestion is a potent conduit to fuel and realize this kind of magic."

The Dark Side of Antidepressants

A longitudinal experiment that followed more than one thousand individuals over a period of forty-five years found that 86 percent of them met the criteria for clinical depression at least once during this period.[78] Much like being suggestible, being occasionally sad, or even depressed, comes with being human. About 280 million people worldwide suffer from depression, which can lead to suicide and other forms of death for more than seven hundred thousand people each year. However, in a study spanning 17.5 million people in the US over a period of eleven years, researchers couldn't find any difference in the quality of life between those who took antidepressants and those who didn't take any antidepressant medications.[79] Moreover, doctors continue to dole out different antidepressants to the masses even when mounting evidence shows that the difference between the best and worst antidepressants is minimal.[80]

Side effects for antidepressants make for a long, extensive list. Symptoms range from impaired sexual function, suboptimal memory, increased risk of dementia, low quality of sleep, digestive problems, and weight gain to increased risk of stroke. Another less-known fact: once started, it's difficult to stop taking antidepressants; the brain adapts to the drug. As a result, many people end up taking antidepressants for life. The way to wean off these drugs requires gradually reducing the dose, with a time course spanning months and sometimes years, compared to the gradual reduction over several weeks that existing recommendations describe.[81]

And yet, not everybody follows this advice, and of the people who try to just quit taking antidepressants "cold turkey," about 30–50 percent fail to do so. In one study, 55 percent of people who tried to quit reported physical withdrawal symptoms, and 27 percent said they had an addiction to antidepressants.[82] Even worse, not only do 50–80 percent of those on antidepressants suffer from sexual dysfunction, but libido problems can linger even after the treatment has ended—so-called post-SSRI sexual dysfunction (PSSD). In addition to PSSD, after successfully stopping the medication, people who had taken antidepressants developed what is arguably its nastiest side effect—emotional numbness.[83]

And then we need to consider the glaring problem of suicidality, as informed by the warning label appearing on all antidepressant medications. Adolescents and young adults, in particular, run an increased risk of suicidal ideation when taking antidepressants.[84] However, it's not just adolescents and young adults who are at risk; it's all adults. Adult-only data concerning suicides, suicide attempts, antidepressants, and placebos show that antidepressants triple the risk of suicide and quadruple the risk of attempting suicide.[85]

Fortunately, alternatives to antidepressants exist. At the end of this chapter, I list many other treatments that work just as well as, if not better than, antidepressants. These alternatives also have better long-term results.

Why No Change?

When I was younger, I thought that things were relatively simple: I will research, present compelling data, and the medical community will read, understand, and change its ways accordingly. Later, however, I realized that compelling data were not enough; time was also an important ingredient. Specifically, medicine takes a long time to listen to and follow in the footsteps of relevant research findings. Finally, only when my thinning hair turned gray, I realized that the mysterious power of "market forces" can eclipse even the most pointed research findings. So how and when do things change in line with specific research results?[86] And why, despite all the indisputable and growing evidence, do we still see antidepressants as the main form of treatment for depression?

The answer to this complex question is multifactorial. First, medicine has been, is, and will likely continue to be slow to change. We practiced bloodletting for over three thousand years before anyone had the "common" sense to question it. We can credit the sluggish pace to human nature: people avoid disturbing the status quo, even if it's clearly off. Medical types tend to be conservative. Plus, a physician who has liberally prescribed antidepressants for decades will naturally be resistant to the claim that clinically, their effect is on par with that of placebos—after all, physicians know better than to prescribe formal placebos.[87] It's incredibly hard to accept a reality so diametrically opposed to our medical standards of care for depression. No one wants to believe that every day, we give millions of individuals—people with a serious, biological illness that increases suicidality—a treatment based on a backbone drug in modern psychiatry that's as clinically effective as, well, taking a sugar pill. It's going to be hard for anyone to hear these data. In this regard, the reaction we see from physicians constitutes normal human behavior.

And let's not forget the patients, those individuals who actually take these drugs. I personally know multitudes who don't fall into the category of the lazy, over-consuming, morally hazardous people that the health policy literature sometimes profiles.[88] And yet, many are afraid to speak publicly about the inadequacy of their medications for fear of disclosing their diagnoses and of damaging relationships, shattering images, and compromising careers. The antidepressant wars encourage them—and all of us—to reevaluate. Because the debate isn't some theoretical discussion within academic medicine; rather, it's a process that must provide decisive answers to those for whom effectiveness and ineffectiveness are more than research questions—they are life itself, and they involve the science of suggestion.

Arguably, one of the largest hindrances to change is the fact that antidepressants work—placebos just happen to work too. They both work because people are suggestible. And this suggestibility is evident not just when it comes to antidepressants. Consider the vaccine roll-out campaigns during the COVID-19 pandemic. The vaccine may come with a slew of unwanted side effects, including arm soreness, muscle weakness, chills, fever, headaches, aches, and fatigue. Yet, in two-thirds of the vaccinated population, these adverse reactions also occurred in placebo shots, *not* just the vaccine itself.[89] In other words, vaccine side effects may largely be psychogenic, rather than a result of your immune system working in overdrive.

Reports of disagreeable vaccine effects, whether by word of mouth or through newspaper headlines, caused many people to deliberately schedule their vaccination appointments so as to minimize disruption to everyday life. In retrospect, perhaps this strategy wasn't as sound as it seemed; merely expecting that you'll feel unwell after getting a jab can manifest the very symptoms you're trying to avoid. In a similar fashion, people take antidepressants and feel better because they *expect* to feel better. Unfortunately, it's this suggestibility that Big Pharma capitalizes on to sustain its antidepressant market. But this human suggestibility also provides clues to how we should treat depression without resorting to antidepressants.

If you read *The Structure of Scientific Revolutions* by philosopher of science Thomas S. Kuhn,[90] you will discover that he challenges the view that progress in science comes as "development-by-accumulation" of accepted facts and theories. Instead, Kuhn argues that within periods of "quiet normalcy," bouts of "revolutionary science" disrupt the status quo, interrupting common thinking by introducing "anomalies" or by someone throwing a wrench into the process, leading to a new paradigm. Then, these new paradigms begin to pose new questions about the old data,[91] moving things beyond a mere inconvenience (or incongruence) with the previous model,[92] changing the rules of the game,[93] and propelling new research.[94]

Recognizing the power of suggestion has certainly thrown a giant monkey wrench into the world of antidepressants. But according to Kuhn, a paradigm shift is nonrational; rather than propagating through the force of truth and fact, change à la Kuhn follows a mélange of sociology, enthusiasm, and scientific promise. Critics of Kuhn claimed that his model described organized crime as much as it did science.[95] Some opponents even expressed disappointment that Kuhn reduced science—one of the most stunning achievements of our species—to trends in sociology or psychology.[96] But German physicist Max Planck, like Kuhn, also claimed that scientific change doesn't occur because individual scientists change their mind but because successive generations of scientists develop different views.[97] In this regard, a new scientific truth doesn't take hold by convincing the adversaries and making them see the light, but rather because eventually, these opponents fade from the scene, and a new generation replaces both them and their opinions.

Scientific innovation rarely makes its way by gradually winning over and converting the unconvinced. The future belongs to the young if only because of the Planck-inspired aphorism: "Science progresses one funeral at a time." So, even after data-fueled arguments about "The Illusions of Psychiatry,"[98] a feature on *60 Minutes*,[99] and a *Newsweek* cover story,[100] most people still haven't accepted that these popular drugs—antidepressant medications—largely appeal to the suggestible part of our human core.

So, should we prescribe placebos? After all, they're almost as effective as active antidepressant drugs, and they're certainly easier on the side effects and pocketbook. Here's the problem with the placebo scenario: imagine telling a patient that you're going to give them an inactive placebo. They now know that whatever they receive is chemically inert, so they are unlikely to believe that this intervention will help them. After all, common wisdom posits that for a placebo to work, you must think you're taking a real drug. And if they don't believe that it'll work, it most likely won't. Luckily, we don't need to prescribe inactive placebos. Let's list some other treatments that are either comparable to or better than antidepressants, and result in value-added long-term prospects.

What Does Work for Depression?

Foremost, to be absolutely clear, I'm *not* saying that if you are on antidepressants, you should stop taking them. As I've pointed out above, getting off antidepressants can be difficult. You may even need help doing so. Whatever you do, consult your physician first. In addition, if you're on an antidepressant and find it helpful and the side effects tolerable, I'd encourage you to stay the course. The bottom line is that if you are reading this now and think that you may have depression but are now unsure as to where to turn for help, I'd advise you to do *something*.

As we've seen, just doing nothing doesn't help. However, many alternatives to antidepressants exist that are of similar or greater efficacy. Out of the available alternatives, you'd want something that's both safe and at least equally as effective as placebos. Your search process and informed choice are important because research has shown that letting patients decide what kind of treatment they're going to get improves treatment outcomes.[101] To be sure, if I were you, I'd first try things that are safer and less dangerous than antidepressants.

At the beginning of this chapter, I mentioned in passing the documented benefits of physical exercise, acupuncture, omega-3 supplements, tai chi, qigong, and yoga; they all provide some benefits comparable to or better than those of antidepressant medication.[102] But we can press further. For example, whereas short-term studies show almost identical benefits from drug and placebo, long-term data show that psychological interventions, such as cognitive behavioral therapy (CBT), perform significantly better than antidepressant drugs. Some researchers speculate that this advantage of psychotherapy over drugs may stem from evolutionary factors.[103] Whatever the reason, an empirical comparison of CBT with antidepressants showed that following up a year or two after stopping treatment, those who got about twenty weeks of CBT were still doing a lot better than those who got antidepressants.[104] CBT is a suggestion-based therapy with an enduring effect, both in person and over digital platforms.[105] That CBT appears to be as effective as antidepressants in the short term, but more effective in the longer term, is nothing to sneeze at.[106] Similarly, comparing exercise versus antidepressants showed that in the long run, exercise alone does better than a combination of exercise and antidepressants.[107] In other words, albeit counterintuitive, the effects of antidepressants are not additive with either CBT or exercise. Moreover, antidepressants seem to increase the risk of relapse, and adding them to other treatments may produce worse long-term outcomes than nondrug treatment alone.

Psychotherapy and exercise are not the only treatments for depression that hold uplifting potential. Mindfulness, meditation, and other contemplative practices can all help, as well as martial arts, improved nutrition, and better sleep. Moreover, keep in mind that one of the major psychosocial risk factors that predisposes individuals to depression is social isolation. Any activity that involves social interaction, therefore, could help alleviate the symptoms of depression.

Social withdrawal promotes feelings of dissociation not just from our communities but also from ourselves, leading to the loneliness that characterizes depression. Perhaps one reason why antidepressants fail to create any considerable improvement is that they intercede solely at the biological level. By prescribing antidepressants, physicians make little attempt to address issues at the level of personal relationships, social interactions, and friendships, or the lack thereof.

The truth is that the causes, symptoms, and consequences of depression are deeply intertwined with factors that involve more than just pure physiology. In other words, in order to eradicate depression, we likely need to mobilize a treatment model that focuses on the psychosocial causes as much as on its biological underpinnings. However, the biological mechanisms of depression targeted by antidepressants have been repeatedly challenged,[108] paving the road to a larger conversation about psychiatric medications and their effectiveness for psychosocial interventions.[109]

Most hardcore scientists stand behind the claim that many of the alternative treatments for depression are ineffective. For example, homeopathy goes against the core fabric of physics and chemistry. We already know that most homeopathic medications hardly contain even a single effective molecule in brewed concoctions that comprise a hodgepodge of repeatedly diluted and shaken mixtures. However, homeopathy is something that a lot of people *believe in*, and this belief is likely a meaningful undercurrent toward positive effects. In addition, homeopathic medications are largely safe (but not for ailments such as malaria or cancer, please). That's why for the right crowd, homeopathic remedies may well form an effective treatment for depression. Remember, psychology and the social sciences form a science too.[110]

Finally, there's an altogether different type of intervention, which humans have used for more than three thousand years—psychedelics.[111] These substances are known for inducing indescribable sensations, and their intense influence on the brain mirrors their dramatic effects on the human mind.[112] Moreover, such moving experiences often accompany a flood of otherworldly perceptions and feelings: striking visual distortions and disembodied auditory hallucinations, to name a few.[113] Alongside risks and criticisms, these mind-altering substances pose a potentially promising alternative by addressing many of the limitations associated with antidepressants.[114]

Psychedelics for Depression?

Individuals afflicted with depression often fall into pessimistic, distorted patterns of thinking that may further exacerbate their despair—from feeling worthless and perhaps even deserving of their pain to burdening loved ones. Subjective accounts from some individuals who take psychedelics echo the sentiment that the psychoactive experience may open the door to a greater openness to positive interactions and a more optimistic outlook. For example, participants who ingested even a microdose[115]—a fraction of a recreational dose—of LSD experienced strong feelings of happiness, connectedness with others, enhanced emotional empathy, and greater prosocial behavior coupled with a desire to spend time around other people.[116]

Psychedelic substances, such as LSD, may rely on the human ability to increase suggestibility and attribute symbolic efficacy to introspective realizations and emotional insights. During such peak experiences (usually in the context of a psychotherapy session), individuals become unusually amenable to new patterns of thinking, observation, and interpretation that would rarely occur in the absence of such substances.[117]

According to this view, the effects of LSD aren't psychotomimetic (i.e., inducing psychosis), psychotherapeutic, creative, or even spiritual per se, but rather *mind-manifesting*.[118] In other words, LSD seems to act as a reflection of, or magnifying glass onto, our state of mind.[119]

A different type of psychedelic, the substance known as psilocybin— the psychoactive ingredient in magic mushrooms—also bolsters communication between different regions throughout the brain; this enhanced interlinking has the power to elicit notable therapeutic effects.[120] In a study of individuals diagnosed with depression, two doses of psilocybin in tandem with psychotherapy produced large, rapid, and sustained effects countering depression.[121] Over half of the twenty-seven participants (NNT < 2) experienced a clinically significant reduction in depression symptoms, with this response persisting for at least four weeks. So, coupled with psychological support, psilocybin produced major reductions in depression symptoms, even in those with severe treatment-resistant depression.[122]

The therapeutic models by which psychedelic substances and antidepressant medications operate differ fundamentally. Whereas antidepressants operate independently of social parameters and focus on the neurochemical dulling of symptoms, the psychedelic experience relies heavily on set (as in mindset) and setting (as in environmental and contextual factors), which wield an immense influence over the outcome.[123] That's why psychedelics, in concert with supportive therapy, can unglue even some of the most deeply entrenched depressive states by making individuals more receptive to hopeful ideas, narratives, and outlooks.

Enhancing the ability of suggestions to penetrate the mind, psychedelics can help reform the maladaptive thought patterns that reinforce depression, thereby getting at the core issue rather than providing the docile, palliative effects of antidepressants.[124] In this regard, psychedelics may offer an effective alternative to antidepressants; they stand as a hopeful possibility for the future treatment of depression.[125]

But these are early days for the resurgence of psychedelics, and the Holy Grail is still out there. For one, we are uncertain how these substances may affect individuals with, say, cardiac problems or underlying mental disorders, including a tendency to psychosis. For another, research on psychedelics has been sometimes accused of taking more of a New Age than a clinical research study approach, guiding participants toward a wanted outcome.[126] Moreover, these treatments work best with open-minded and willing individuals through a combination of psychedelics in concert with supportive therapy. That's why in reality, wherein psychiatrists spend little time with their patients, largely shy away from "talk" therapy, and focus almost exclusively on a biological model of psychopharmacology, a direct comparison between solo anti-depressants and stand-alone psychedelics greatly misses the mark.[127] But even with such important caveats in mind, being careful to discern empirical findings from spiritual practice, the unrealized potential of these psychoactive substances heralds at least some promise.[128]

Are You Ready for the Revolution?

Although historically, anti-psychedelic smear campaigns led to the criminalization of ecstasy, LSD, and magic mushrooms and drove most clinicians and researchers away from them, we are beginning to witness a comeback of the old psychedelics. Not only is the momentum palpable with popular publications in top medical journals—for example, on the benefits of treating depression with psilocybin[129]—but it also manifests with the first phase 3 clinical trial showing that MDMA (ecstasy/molly) paired with psychological counseling brought marked relief to patients with severe post-traumatic stress disorder (PTSD).[130] In fact, MDMA had started out as a dissociative therapy drug before it hit the party scenes and got stigmatized. The timing and cumulative effect of this body of work excites scientists, practitioners, and entrepreneurs in the rapidly expanding field of psychedelic psychiatry, as well as psychedelic medicine. It seems inevitable that the FDA should grant approval for psychoactive compounds for therapeutic use—with MDMA and psilocybin as prime candidates. Notably, psilocybin seems to be one of the safer drugs in psychiatry.[131]

Even with reservations lingering,[132] psychedelics hold profound implications for clinical psychiatry, a field that over recent decades has seen too few pharmacological advancements for the treatment of mental disorders and addictions. From pop books[133] to public talks,[134] the changing social climate fosters a destigmatization of psychedelics. Moreover, against a backdrop of a mental health crisis and a general disillusionment with backbone psychiatric medications—antidepressants—the nascent potential of psychedelic research seems to fill the void with optimism and recognition that psychiatry is itching for new therapeutic tools in its clinical toolbox.

The landscape has become less barren with organizations such as the Multidisciplinary Association for Psychedelic Studies (MAPS)—a multimillion-dollar research and advocacy group that employs more than one hundred neuroscientists, pharmacologists, and regulatory specialists—laying the groundwork for the looming psychedelic coup. Top universities are racing to set up psychedelic research centers. For example, Johns Hopkins, Yale, the University of California, Berkeley, and Mount Sinai Hospital in New York are among the institutions that have recently established psychedelic research divisions, with generous financing from private donors.

Suddenly, psychedelics are awash with money as investors keep pouring millions into start-ups. While Big Pharma typically runs conservative research, and insurance companies require ample convincing to cover psychedelic-assisted treatments, the enthusiasm for the transformative power of psychedelics is spawning new initiatives independent of these sluggish mammoths. More than a dozen new companies have appeared, with a handful that have gone public valued at more than $2 billion collectively.[135] As a case in point, Field Trip Health, a young Canadian company that trades on the New York Stock Exchange, has raised $150 million to finance dozens of high-end studies on ketamine— a dissociative anesthetic that has some hallucinogenic effects—in clinics around Los Angeles, Chicago, Houston, and other cities across North America. Compass Pathways, a Nasdaq-listed health-care company that has raised $240 million, is conducting two dozen clinical trials across ten countries of psilocybin therapy for treatment-resistant depression.[136]

Critics have expressed concern about undue financial and ethical influence within this sometimes mystically infused research field.[137] Indeed, psychedelics induce mystical experiences, which may in turn help improve conditions such as depression, anxiety, addiction, and end-of-life terror. Psychonauts, proponents of psychedelics who advocate to integrate them into mainstream society, want to unlock their transformative power for humanity. However, some opponents claim that psychonauts influence public opinion in favor of psychedelics using the pretense of rigorous research, bestowing the imprimatur of science in name only.[138]

As we emerge from the COVID-19 pandemic and continue to fight fentanyl and the opioid epidemic, an incipient mental health tsunami is now slamming onto the shores of our already-burdened health system. The combination of suggestion science and psychedelic interventions may herald the next revolution in psychiatry.[139] Moreover, outside of medicine, it may signal a forthcoming global revolution in the way humans connect with one another at a social and cultural level.[140] This trajectory raises overarching questions about our ability to leverage ripples in suggestibility profiles and harness suggestion-enhancing techniques not just in the contexts of depression or medical treatment, but within domains pertaining to our global fabric of society and culture.[141]

CHAPTER 8

A Nexus of Psychedelics, Suggestion, Society, and Culture

I maging techniques of the living human brain reveal that psychedelic substances, such as LSD, generate a spectacular display of fireworks in the tissue between our ears. Regions of the brain that are otherwise "quiet" suddenly ramp up and bustle with activity.[1] This boost in neural commotion creates new forms of neurobiological interconnectivity, joining and linking up different cortical areas. In this way, psychedelics may accelerate a process of neuroplasticity—the idea that the brain can change and reform itself by building and crafting new neural pathways.[2] This dynamic fuels perceptual and behavioral changes and, importantly, alterations in the interpretation of meaning.[3]

This "meaning effect"[4] describes the ability to mold, shape, and reassign the value we attribute to things. Even for rational, skeptical, and deeply nonspiritual folks, psychedelic experiences sometimes carry transcendental or mystical overtones bordering on metaphysical or religious revelation; they can change attitudes and beliefs![5] The ability to change core human identity components, such as attitudes and beliefs, is not just of medical importance; it carries global implications for the way we live, interact, and cooperate both within our communities and transculturally.[6]

States and cities are beginning to loosen regulatory restrictions, and the advent of federal decriminalization of psychedelics for therapeutic and recreational use is already within sight. Propelled by the nation's accelerating embrace of recreational marijuana and by public weariness over an endless war on drugs, Oregon became the first state to legalize the therapeutic use of psilocybin. Denver, Colorado; Oakland, California; and Washington, DC, have decriminalized psilocybin, and several states, including California, are moving along with similar legislation. Even the staunchest of conservative legislatures are investing in psychedelic research. For example, as of May 2021, the Texas Legislature passed a bill allocating millions of dollars to the Houston Veterans Affairs Medical Center to study psilocybin for PTSD.[7] Although the drugs remain illegal under federal law, the Justice Department has so far taken a hands-off approach to enforcement, similar to how it has handled recreational marijuana.

Classic psychedelics, including LSD and psilocybin, are neither addictive nor causes of organ damage, even in mega doses. But how far and how fast the pendulum should swing in their favor is more than a question about a fashion trend to commercialize psychedelics, slacken social prohibitions, or advocate for adult recreational use. Rather, it's about a judicious call for the application of science—clinical-medical, computational, life, but also social, anthropological, and psychological science. Suggestion science is part and parcel of this package. For example, a professorship fund with a $20 million funding goal at Johns Hopkins University aims to examine how psychedelic experiences relate to aspects of spirituality, worldview, well-being, and prosocial behavior.[8]

With one of their primary reparative hallmarks being ego dissolution—a reduction in the sense of being a self or an autonomous "I" distinct from the rest of the world—psychedelic substances tend to melt resistance and rigidity and bolster openness and adaptability. While ego dissolution compromises our sense of "self,"[9] and has the potential to bump some individuals "out of orbit" by manifesting negative outcomes, it seems to render many others more agreeable to accepting and addressing helpful suggestions.[10]

In this regard, we can increase suggestibility not just through psychedelics but also through extra-pharmacological factors such as food, music, lighting, aromas, general atmosphere, and social interaction with other people. For example, nondrugged users can experience drug-like effects from exposure to users experiencing a particular drug.[11] In my lab, my students and I demonstrated that some people ingesting a placebo can experience hallucinatory-like effects when the setting suggests a psychedelic party.[12] In other words, we were able to show experimentally that suggestion and expectation can play a major role not just individually but also in group and social settings. We will get back to this idea of social contagion later in this chapter.

In classic social psychology experiments, such as Solomon Asch's conformity studies from the 1950s, results typically indicate that an individual can change personal beliefs, behaviors, and reasoning in response to social pressures. Asch found, for example, that about 75 percent of participants conformed to a group's overtly incorrect answers at least once in order to avoid social ridicule.[13] We have come a long way since the 1950s, further elucidating the influence of suggestion within a group context.[14] Moreover, studies show that such group dynamics often give rise to positive effects on social interactions and interpersonal relationships because users report greater sentiments of connectedness and belonging within the group.[15] This group-driven increase in suggestibility may be in part due to the shared, profound experience that enriches social bonds.[16] Accordingly, psychological, community, and social support systems make for an essential component of any curative psychedelic experience.

Psychedelics and Suggestion

Because the psychedelic experience tickles our suggestibility, it opens the door to a bigger discussion about how the suggestions and the support that accompany taking psychedelic substances influence our behavior. Researchers show that we can substantially increase our suggestibility during the psychedelic experience by managing a host of nonpharmacological factors.[17] As mentioned earlier, mindset—including expectations, perceptions, and psychocultural beliefs, along with the physical environment surrounding the intake experience—is well-documented to influence outcomes in psychedelic research,[18] so much so that these factors now have a special term of art: "set and setting."[19]

When it comes to therapy and clinical interventions, set and setting can influence health outcomes by bolstering suggestibility and paving the way for healing suggestions that could alter physiology in a desirable direction. That's why trained support and professional monitoring form essential ingredients of the psychedelic experience. This type of supervised reassurance makes a difference in minimizing unsought abreactions—the release of a previously repressed emotion through reliving the experience that caused it—and maximizing the desired outcome. On the flip side, heightened suggestibility comes with its inherent perils: individuals who use psychedelics may fall victim to potentially wayward suggestions. In the presence of the wrong company and ideas, psychedelic users may be more susceptible to believing all sorts of misguided notions—from disturbing false memories to outlandish conspiracy theories.[20]

Futzing around with suggestibility is a double-edged sword. Psychedelics have the potential to leverage the power of suggestion for good: alleviate suffering, enhance personal growth, and promote well-being. But if used indiscriminately and injudiciously, these substances have the potential to precipitate a wave of irrational thinking and further blur the lines between truth and "truthiness"—something that feels true but isn't.[21]

Suggestion and Society/Culture

In Chapter 2 we saw that as a species, we are still evolving;[22] however, we now know that our evolutionary process is not just about biology, genetics, and DNA, but also about culture, demographics, and technology.[23] Modern humans, in particular, have recently opened two Pandora's boxes—that of climate change and that of artificial intelligence—and unleashed environmental and technological perturbations that may dramatically influence our future evolution.[24]

Culture works faster than genes and helps us with niche construction. In this way, suggestion and suggestibility likely provide an evolutionary advantage. Our higher brain functions are distinctive—for example, we are conscious of our consciousness and aware of our awareness. Moreover, human language appears to be a distinguishing phenomenon, without meaningful analogues in the animal world.[25] Our linguistic prowess involves both representational gestures and language but mostly relies on collaborative computation, a process unique to humans.[26] These qualities allow us to engage with collective imagination, social identity, and norm psychology—the cognitive tools needed for cooperation.[27] Moreover, they provide us with the ability to pretend and spend our mental energy on what others may be thinking and how we can communicate with them, let alone influence their thoughts and behaviors. Suggestion is key to this ability.

But suggestions also entail a potential for societal peril, mental abuse, and gaslighting. A common occurrence, for example, is that of convincing a sexual assault victim through suggestion that their experience was less serious than they remember. Studies estimate that two out of three sexual assaults go unreported.[28] In the US, only 18 percent of reported rapes lead to an arrest, and 2 percent result in a conviction. Part of the reason for this awful situation hinges on the heartrending discrepancy between the evidence that the criminal justice system wants to see and the nature of traumatic memories.[29] This incongruity further undermines the ability of jurors to objectively assess court cases.[30]

Traumatic experiences shake up our memories to a point where we may misremember what exactly happened. Even from a distance of decades, Holocaust survivors and many people who have been violated, raped, or sexually assaulted often retain vivid memories of specific details associated with their experience, such as smells and sounds. And yet, upon interrogation, they cannot recall exactly who and what was where and at what time—information that law enforcement and officers of the court often require to establish the facts of a crime. Subsequently, these victims may contradict themselves, thereby undermining their own testimony.

The neural encoding of traumatic events follows a different process from common, everyday memories. During trauma, the human brain secretes stress hormones—for example, cortisol and adrenalin—which direct us to attend to details of the "here" and "now" instead of the overarching, bird's-eye view of an event. When under attack, it makes more evolutionary sense to focus on what we are experiencing than to interpret global meaning. But for purposes of evidentiary testimony in court, fractional recall of selective details may constitute an incoherent account or an unreliable witness, and may well compromise the likelihood of a conviction.[31] This is another unfortunate example where psychological phenomena are incongruent with and misunderstood by the legal system. Unfortunately, it's not entirely possible to avoid these types of situations without changing legislation and the rules of evidence.

Sidestepping the thorny issue of traumatic memories, certain metacognitive and mental workouts can help us use suggestion and mindset to develop and maintain individual resilience. In that vein, here is a helpful exercise you can do to drill and explore your evolutionary gifts of suggestion. Imagine that your thoughts are available for everyone to see. In other words, you go about your day while whatever you are thinking between your ears becomes instantly transparent and immediately accessible to others.[32] If you think that a particular person is an obnoxious scumbag or sexually attractive, they can tell what you're thinking. It's a tricky cognitive exercise, one that's good for getting in touch with your inner thoughts and perhaps your suggestible self. If you try it, you will soon discover that this type of brain training is difficult but worthwhile, even in short bursts and over modest intervals.[33] It leads us to practice a form of self-administered cognitive behavioral therapy (CBT) wherein we learn to "catch" or identify undesirable thoughts by gaining an awareness of our thinking process, "challenge" these thoughts by reflecting on what makes us think this way, and then "change" our thoughts. These three Cs have long served as a pillar of effective CBT. Moreover, we can extend this type of suggestion from the level of a single individual to that of a group.

Mass Suggestion and Social Contagion

If you think that influencing the behavior of one person in line with a suggestion is cool, imagine doing the same to a collection of individuals.

I've always been fascinated by and attracted to the dynamic of cliques, communities, and groups. I even married a woman who wrote her senior thesis at Harvard on the mass suggestion of the Salem witch trials (Massachusetts, 1692–1693), the deadliest witch hunt in colonial North America, with more than two hundred people accused of practicing witchcraft and twenty-five sentenced to death. Although those trials took place centuries ago, many things have (and haven't) come a long way since.

Examples of mass suggestion—for example, cases when psychological conflict and distress convert into aches and pains with no physical origin—date back at least to European convents of the late fourteenth century. Nuns would sometimes display vigorous tics, foam at the mouth, express themselves obscenely, climb trees, and even meow like cats. Obsessed with sin and immorality, and deeply embedded in a landscape laden with mystical supernaturalist thinking, they constantly monitored their own thoughts for the slightest sign of moral transgression. Those who found "impure" thoughts believed that demons had penetrated their souls, and their behavior mirrored what exorcists—and the religious—expected of the diabolically possessed. Often, through social contagion, such behavior would also "bleed out" and spread to witnesses who shared the same religious belief.[34]

Before the introduction of social media, contagion occurred when a group shared a common physical space for a substantial period—hours on airplanes, long days in school, weeks over summer camp, months in specialty training, years in prison. Nowadays, one doesn't even need to be in physical proximity to "catch" such symptoms. Ready access to the internet will take care of that... If, previously, priests, rabbis, imams, or other clergy exorcised demons, dybbuks, jinns, or other spiritual entities in person, now smartphones, messaging services, and digital social networks are the new go-to locations of social contagion.

We have replaced physical gatherings, where people come together to share common beliefs, with virtual spaces where people engage in the same collective behavior. In this regard, whereas geography largely limited the epidemic hysterias of yore, today our globally connected world magnifies such impact, instantly scattering information over vast distances and territories. Belief serves as fuel; technology as the vehicle of dissemination. Virtual spaces have become our houses of worship and pulpits.

Even before the explosion of social media, the US had documented cases of mass suggestion. For example, in early 1990, thirty-four toll workers on the Triborough Bridge connecting Manhattan, Queens, and the Bronx in New York City reported sudden, flu-like symptoms. After a rigorous investigation by experts in epidemiology, environmental contamination, and occupational health, all signs pointed to the conclusion that the symptoms were likely psychosomatic in nature—a case of mass suggestion. However, their employer was reluctant to endorse the medical diagnosis of "it's all in your heads" because the employees were really feeling the symptoms, and they were not crazy.

A spokeswoman for the Triborough Bridge and Tunnel Authority (TBTA) said, "No, I don't think we want to prove it as hysteria. These people work very hard, and we don't want to make light of what happened to them."[35] In other words, the TBTA leadership did not want its employees to feel disparaged and unsupported. The moral and ethical conundrum in these situations brings into focus the juxtaposition of the diagnosis of mass suggestion and the erroneous dismissal of the symptoms experienced by the afflicted as illusory or not real.

Moreover, treating mass psychogenic cases of this sort is a double-edged sword. On the one hand, a clear public announcement that some symptoms are mostly psychological is necessary to begin an effective intervention (even though also likely to facilitate their dissemination). On the other hand, to label individuals as "hysterical" will probably alienate them and their families. We must achieve a delicate balance, therefore: to convey our scientific understanding without appearing accusatory or trivializing the plight of those afflicted. But all this took place before the first recognizable social media platform appeared in 1997.

One of the first cases of US-based mass suggestion in the social media era occurred in 2011. High school students in Le Roy, New York (near the Canadian border), broke out in verbal outbursts, tics, seizures, and speech difficulties.[36] One girl with Tourette-like symptoms likely "spread" them to more than a dozen other students. The effects felt real, but their causes were not physical in nature; the students were subconsciously converting stress into physical symptoms.

"It's a very hard pill for me to swallow—what are we, living in the 1600s?" the guardian of one of the students said. No, we are living in the twenty-first century, but mass suggestion, fueled by digital social networks, thrives on communicated anxiety—it flourishes on a collective stress response. Doctors witnessed an increase in cases in proportion to the sensational media coverage and online discussions. As more students got sick, the story got bigger, and then even more students got sick.

Before long, the community of Le Roy was distinguishing between kids who were "genuinely sick," those whose illness was "psychological," and those who were "faking it" so they could get on TV. No matter how many times the doctors explained that these symptoms were real, something the girls could not control, the finger-pointing persisted. The accusations, thrust wildly over internet forums and social media, invariably exacerbated the symptoms. If you have to prove that you are ill, you can't get better.[37]

In the final analysis, certain trends emerged: the way the symptoms seemed to flow from the students, mostly girls, at the top of the social heap to those who looked up to them, the commonality of a certain kind of background vulnerability—for example, abuse and troubling family circumstances—and the unique social dynamic between cheer-leading, art class, and female friendship. But the workings of these mass-suggestive phenomena remain, in many ways, as enigmatic as adolescence itself.

As a case in point, during the COVID-19 pandemic, social media consumption increased greatly, especially among adolescents, along-side an upsurge in motor twitching and functional tic-like behavior. Because many of the video clips floating around social media repre-sent misleading and false information about movement disorders, they encourage and reinforce irregular motor expression and likely fuel the rapid increase in functional tic-like behavior. Such accounts typically come about as a result of exposure to social media but also of long-term stress, which may bring about neurological disruptions. When they occur, such symptoms may include not just motor twitching and shaking, but also facial and vocal tics, garbled speech, and trance-like acts. This odd behavior goes by many nuanced names, including mass psychogenic factors, conversion disorder, and mass suggestion. These symptoms tend to occur mostly in teenagers, and more so among teen-age girls.[38]

By some eerie stroke of historical irony, about three hundred years after the Salem witch trials, a variation resurfaced in the same locale—first in Ipswich (2004) and later around Danvers (2013), Massachu-setts, the latter a town once known as "Old Salem" and "Salem Village." Two dozen students—mostly girls at the Essex Agricultural and Technical School in Danvers—displayed monthslong "mysterious hiccups" and vocal tics.[39]

Even without a crystal ball, it seems reasonable to expect that mass suggestion will continue to haunt our species. Such phenomena will likely periodically surface and mirror the knowledge and events of our contemporary culture and social dynamic. Although it's difficult to predict what appearance it will assume and when it will emerge, such collective psychogenic eruptions will likely feed on our fear, anxiety, and uncertainty. In the same way that we are now afraid of invisible viruses, environmental threats, and terrorist attacks, so will we dread other things—perhaps futuristic weapons, advanced technology, and misinformation.[40]

Trust Me, It's Fake News

Once, in my teenage adventures around Tel Aviv, I happened upon a fluke discovery. As I was walking to a specific address in the city, I noticed that the house numbers on my side of the street were out of sequence. It was no biggie, just a small jump from 47 to 51 and skipping 49, but I made a mental note of it. Over the years, I collected more and more examples of these kinds of sequence interruptions in house numberings across assorted metropolises. I even recorded these freak address disruptions in a special notebook. I found these aberrations fascinating, and it got me thinking.

Before long, I plotted a new slapdash psychological experiment. Every time I took a taxi—we are talking about a time before Uber and Lyft—to or from one of my lectures or magic shows, I'd engage the driver in a friendly conversation about magic. When the opportune moment came about, I'd casually mention that I once "vanished a building but couldn't bring it back." Most drivers were skeptical and perhaps thought I was a nutjob, but occasionally I'd get a cabby who was keen on hearing more. I'd make sure to mention the street address of the missing house, knowing that at least some of them would be curious enough to check it out. I also knew that many of these drivers were yappers: they liked to make conversation to pass the time while driving, and I counted on them to disseminate my story far and wide within the community.

Sure enough, word got around. One day, I heard a story about a "paranormalist" who vanished a house at the address that I'd indicated. What's more, the driver who related the story to me—someone I've never met before—swore that he was present at the show and saw it with his own eyes. Any good game of broken telephone will end with a completely different story than it began with, and I was pleased to see how my original scheme developed so nicely. Another driver told me about a disgruntled "sorcerer" who vanished an entire building because he didn't get paid for his performance. It was duly amusing, but the tale came with an address I hadn't heard of before—it wasn't in my little notebook. Just as curious as the cabbies I'd duped, I headed to that address and discovered a perfectly erect house in good standing. I suppose I must have restored the vanished building after finally receiving payment. Even powerful magicians cannot remember well; as we discovered in Chapter 6, memory is unreliable.

Consider another example, an experiment that examined the issue of abortion in the Republic of Ireland. This is an issue that prompts divisive discourse within the public;[41] moreover, Ireland used to host some of the most restrictive abortion laws in the world, mostly because it has been a primarily Catholic country and is against it based on religious beliefs.[42] Researchers asked participants to look over short news stories of campaign events, accompanied by a photograph, and to tell them how well they remembered the event, where they had heard of it, and how they had felt about it. Some of these events were true items that really happened, but some were specious, baseless, and fabricated.

Looking for false memories in the week preceding the 2018 Irish abortion referendum, about half of the study's three thousand participants reported a false memory for at least one of two fabricated scandals that researchers had circulated, with more than one-third of participants reporting a specific memory of that made-up event. As you may expect, voters in favor of legalizing abortion were more likely to "remember" a fabricated scandal regarding the opposing campaign, and vice versa.[43] In other words, people suggest themselves into remembering things that didn't happen when it serves their belief system.

Another interesting bit is that suggesting to people that they may have been exposed to fake news did very little to reduce the rates of false memories and did not eliminate this effect of bias, with participants less likely to detect fake news that was in line with their beliefs. As we have already seen, people are easily suggestible when it comes to party lines.

In politics, fake information has become the main currency of our time; politicians circulate all sorts of communications to get us to vote for them—and many people actually believe them. Naturally, we are more susceptible to suggestions that reaffirm beliefs we already hold. So when researchers presented doctored photographs of President Obama shaking hands with the president of Iran as a news item, Republicans were predisposed to accept this information. The same is true on the other side of the aisle: Democrats are more likely to "remember" the time President Bush vacationed with a professional baseball player during the Hurricane Katrina disaster.[44]

Although both of these fabricated stories never happened, this adherence to party lines stands out as a robust human tendency that caters to our suggestible brain. In this regard, perhaps what has changed isn't the fact that modern politicians lie differently than they did before, but that we are now more amenable to and more accepting of their suggestive narratives.

*   *   *

Only a few short decades ago, three main news channels dominated American TVs: ABC, CBS, and NBC. Whether you consumed your news from one or the other two made little difference; they shared the same ideological leanings, and the information they communicated was largely the same. Today, on the other hand, the situation has turned on its head: we have a huge selection of news sources to choose from, each promoting explicit ideas and sometimes ulterior agendas, and people listen to what they want to hear. Subsequently, in our modern world of multisource media, we can no longer agree on the facts, we cannot settle on what constitutes reality, and we have little consensus about truth and trust.

Truthiness—not truth itself but the truth we'd like to be true, the truth that feels good and right—was word of the year for 2005 (by the American Dialect Society) and 2006 (by Merriam-Webster). Who can tell me that George Washington was a slaveholder if I want to believe that he wasn't? It feels powerful to turn factoids into concepts we wish or believe to be true, rather than follow bona fide facts. Are you concerned about the nearly extinct northern white rhinoceros? You can "save" them by rewriting their population number on Wikipedia. This form of Wikiality brings "democracy" to information.

The ascent of parallel Wikipedias around the world epitomizes the dawn of relative truth. Wikipedia is perhaps the first encyclopedia in the history of humanity that contains explicit inconsistencies and incongruences relative to differing human experiences. For example, among the fifty million entries in hundreds of languages, if you compare the entry for "Jerusalem" in the Arabic and Hebrew versions, you'll find different content. The longer and more detailed Arabic version focuses on the city's history, culture, and demographics as well as its religious significance to Islam and Christianity, whereas the Hebrew version focuses on the city's religious significance to Judaism. In addition, some ideological communities adopt the Wikipedic model and create their own versions—for example, by excluding references to evolution—thereby penetrating different social spheres and organizing knowledge according to the beliefs and perspectives of specific sectors.[45]

In the US, this trend has set new records, with about one-third of the population being of the opinion that President Biden was elected illegitimately; millions of people have become convinced that the 2020 elections were fake. That's both astonishing and scary.[46] Moreover, social polarization and the fact that people live in separate "worlds of knowledge" within the same country make it very difficult to believe in and follow a single leader. In the US, specific leaders—through their personalities and actions—have spawned tremendous social disunity. Even more disturbing, social media platforms now add a twist to this fabric: if you put "climate change" in your search, for example, you will get different answers based on whether you are connecting from the Northern or the Southern US states. In other words, you will receive different information on whether or not climate change is a "hoax" based on your location![47]

Fake news has an economic advantage because lies work better than truth on the web: sensationalism and clickbait sell. Specific communities engage with fake news sources, and publishers with weak journalistic standards typically produce misinformation—false information devoid of the intention to deceive or mislead—and downright disinformation, a deliberate, intentional effort to mislead by spreading false information.[48] Once produced, fake news stories diffuse farther, faster, deeper, and more broadly than the truth.[49] Whether deliberately or unwittingly, humans—potentiated by suggestibility, exposure to social media, and, more recently, exposure to artificial intelligence applications—are increasingly losing their ability to tell the difference between what's true and what *seems* true.

Today, we know that social media has caused a lot of damage, for example by weakening the fabric of societal consensus. Online conspiracies have increased the number of Holocaust deniers, anti-vaxxers, flat-earthers, and other theorizers who consume alternative scientific and historical ideas. People immerse themselves in online communities that tell them what they want to hear. In this way, they create Petri dishes for complementary worlds to thrive without interruption. Group recommendation algorithms further abet the spread of misinformation.[50]

Living in a post-truth era is challenging. No matter how careful you are, and what precautions you take, fake news will get you. In addition, deepfake—the ability to easily and cheaply generate convincing images, audio, and video via artificial intelligence applications and through machine learning—is now on the rise.[51] For many Americans, let alone for global citizens of the world, the combination of fake news, alternative facts, and deepfakes has called into question how we construe reality itself. We need to have the touch of a magic illusionist and a dab of a research scientist to navigate our confusing world.

Suggestion and Social Justice

Can we leverage research findings about suggestion to advance social causes and promote agendas concerning race, gender, and topical issues related to our cultural mosaic? I think we can, but I think it requires some serious commitment by our political, community, and science leaders, perhaps peppered with a smidgeon of how magicians think.[52]

More than four hundred years after the first slave ship arrived in America, can suggestion help correct racial inequalities, bigotry, and hatred? After all, the police are more likely to kill people with mental disabilities, especially disabled people of color, than their neurotypical counterparts;[53] hospitals are more likely to report Black and Latino families for child abuse, and more likely to avoid reporting comparable situations in white families;[54] practitioners of mental health are more likely to misdiagnose African American patients with major depressive disorder as having schizophrenia, in part due to a racial bias,[55] and the unconscious bias of physicians may further contribute to racial and ethnic disparities in medical procedures.[56] Can suggestions make a difference?

Social psychologists have long studied how malleable our automatic attitudes are and whether we can counteract such involuntary prejudice. For example, researchers found that exposure to pictures of admired Black and disliked white individuals reduced implicit preference for white over Black Americans. In other words, by repeatedly showing participants photographs of famous and respected Black people, such as Martin Luther King Jr., and photographs of infamous and disapproved white people, such as criminal and cult leader Charles Manson, negative attitudes waned by more than 50 percent.[57] Similarly, videos of Black people during positive activities such as going to church or enjoying a family barbecue also reduced implicit bias.[58] These early efforts to change the social context and, through it, reduce automatic prejudice and preference have paved the road to diversity education.[59]

Today, regulators and diversity and equity initiatives in the workplace require teachers, physicians, managers, and employees to undergo cultural sensitivity training. But such forcing may result in a backlash because of reactance—the need to preserve psychological autonomy.[60] People who go through such mandatory interventions sometimes perceive the training as a threat to their freedom of expression, or they simply take offense at the intimation that they may entertain a prejudice in the first place.

Although nearly all colleges and universities, as well as most workplaces in America, conduct diversity training workshops, few institutions actually care to assess and evaluate their effectiveness. Research findings suggest that pressure to conform to minority-focused standards, or similar cultural benchmarks, rarely yields uniformly positive effects, and may actually backfire.[61] In other words, these training programs seldom work well and may even make things worse.

What's the alternative?[62] People tend to suppress racial prejudice when feedback indicates that many others disagree with their racially prejudiced stance. In this way, providing consensus information may be a more effective way to change racial beliefs. We know that the opinions of in-group members carry more weight than those of out-group folks, so learning about the racial beliefs of others has the potential to either produce or inhibit stereotype change.[63] Racial stereotypes often persist because people assume that their stereotypic beliefs are congruent with those of others, and often overestimate the negative stereotypes that others uphold. Suggestion can undermine negative stereotypes through the presentation of countervailing consensus feedback. Withdrawing the social backing from an idea goes a long way toward undermining the power of that idea over an individual. If a colleague cracks a racist joke or an ethnic slur, call them out! That's one of the more effective ways of countering racism.

My students and I have developed illusion-based experiments[64] that draw, for example, on non-Black participants exchanging bodies—in other words, "body swapping"—with a Black out-group member using virtual reality.[65] Looking at yourself in the mirror and perceiving yourself as having a different skin color through the use of virtual reality technology leads our brain to feel that we are now pigmented differently through an active, transformative experience with technology.[66] Such procedures increase self-other merging as well as empathy, and improve intergroup relations. Experiments that rely on variables that change self-perception—for example, when participants watch their avatars, digital versions of themselves, playing tennis, and gradually getting thinner from the exertion—wield physiological effects on real people.[67] These types of interventions represent a creative new direction to illuminate problems of racial and gender tension, and how we may go about addressing them. In this way, and others, we can apply the science of suggestion—in a way that transcends lip service and performative activism—to result in real social impact, health outcomes, and political change. We already saw that suggestion can make you lose weight and change your physiology, but it seems we are on the cusp of something bigger: extending this call to action and making the world a better place.

Suggestion, Magicians, and Scientists

Whether personal or communal, narratives are powerful because our thoughts and perspectives can contribute to and shape our well-being. Moreover, expectations, beliefs, and mindsets influence how our bodies function, how we heal, and even how we age.[68] In my lab-based studies—although drawing on much smaller samples and less ecological conditions than in the general population—my team was able to demonstrate that suggestion can affect the subjective experience of even something as biological as visual acuity[69] through psychological-ophthalmological interactions,[70] or by influencing our mindset.[71] Other researchers have shown similar results; for example, if you rig standard eye charts so that participants will identify some of the smaller-print letters correctly from the beginning of the exam, giving them the expectation that they actually see better, they improve their overall eye test scores.[72] Similarly, wounds heal faster when participants stay in the presence of accelerated clocks; when you think that time is passing faster, your body heals faster! On the flip side, discouraging health news can contribute to worsening effects: learning that you are prediabetic, even if you surpass the threshold by a hair, may actually play a part in the development of diabetes. Clearly, our state of mind holds some sway over our health and behavior.[73]

Some mental illnesses correlate strongly with forms of suggestibility.[74] Similarly, neurological and psychiatric conditions have been associated with aspects of creativity.[75] For example, the 2001 film *A Beautiful Mind* tells the story of mathematician John Nash and how schizophrenia affected his creative genius; actor and comedian Robin Williams had well-documented struggles with bipolar disorder, alcoholism, and Lewy body dementia; and composer Dmitri Shostakovich carried a piece of shrapnel in his brain for the last thirty-four years of his life and used it to help him compose music. Each time he leaned his head to one side he could hear music, and his conscious mind was filled with melodies that guided his composing. Of his fifteen quartets, arguably his most powerful work, he composed all but one after this German fragment launched into his brain in 1941.[76]

Moreover, some researchers argue that psychopathology—from anxiety or depression to personality disorders or psychosis—and neurodivergence (e.g., autism spectrum disorder and attention deficit hyperactivity disorder) can enhance creative thinking.[77] In this regard, mind-wandering and disorganized thinking may pose a challenge for sustained attention but seem advantageous for creative outlets.[78]

And yet, among creative performers, magicians seem different. They fluctuate widely in their creativity: few are innovative trailblazers, whereas most lead successful careers by performing old-hat tricks, sometimes with their own tweaks, without the need to invent new magic effects. In addition, the magician's oath not to share their secrets allows them to perform the same tricks repeatedly while maintaining an air of mystery.

Like comedians, magicians are unique in that they also craft and perform their own shows; most other creative groups either create or perform, but not both. But unlike comedians, magicians risk a whole lot more. For example, a lame joke by a comedian makes for an unpleasant, but hardly destructive, blow to a comedy act; however, one botched magic trick will destroy a magic performance with little opportunity for redemption. Magicians, therefore, must exercise continuous vigilance while entertaining their audience.

Researchers surveyed nearly two hundred magicians with an average of thirty-five years of experience in performing magic compared to a sample of nonmagicians—including comedians, poets, actors, and musicians—with a similar age range and gender distribution. They found that, on the one hand, magicians did not exhibit any predisposition for autistic traits, scoring similarly to the general population. On the other hand, magicians scored lower on nearly every psychotic symptom compared to the general sample and other creative groups: a very high ability to concentrate, lower levels of social anxiety, and fewer instances of unusual experiences, distorted thoughts, and hallucinations. In a way, the magicians were more like scientists, who also score low on psychotic symptoms.[79]

We have come full circle. A journey from magician to scientist—through the land of suggestion, with stops along the way to glean insights from hypnosis, evolution, psychological and brain science, and the worlds of research and mental health, atop a rapprochement of suggestion-enhancing psychedelics within our fast-moving culture—ends with some parallel reverberations between magicians and scientists.

As we saw throughout this book, both scientists and magicians require high levels of organization and perseverance in their work. Moreover, just as scientists often consider and explore different solutions to the same problem, so can magicians perform the same magic trick using multiple methods. When I perform a cool trick in front of my science colleagues, they often ask, in the same way most children will, that I do it again so they can both marvel at and better study the effect more closely. I usually don't; magicians typically perform a trick only once. But sometimes, if the vibe is right, I will indulge them with a repeat, sometimes more than one.

What scientists don't always appreciate, however, despite plentiful disclaimers, is that a good magician can sometimes conjure up the same effect in different ways. For example, when I bend a spoon "with my psychic powers," they ask me to do it again, only this time with my sleeves rolled up. I reluctantly do so. But when they ask that I do it a third time, I can do it with my sleeves down again because, congruent with the scientist's mindset, "I have already demonstrated that the sleeves have nothing to do with it." Little do they know, however, that when I roll them back down, I could—if I wanted to—use a different technique and have something up my sleeve!

This magician-scientist synergy has served as a personal and professional theme in my life, as well as throughout this book. People often ask, tongue in cheek, if I am the best magician among the scientists or the best scientist among the magicians. I answer that I am probably neither but that the ability to draw on and benefit from both worlds, including the unique techniques, skills, and systems of knowledge related to each, has made my journey into the suggestible brain all the richer and more meaningful. I hope you've enjoyed the ride as much as I have.

CONCLUSION

Parting Suggestions

E ccentric" and "out there" used to describe research on the science of suggestion; however, this domain has begun to gain traction over the past decade or so. This budding acceptance is less a result of how research has changed and more of a testament to how our world has changed. Today, mind-body effects have become a mainstream topic. More people acknowledge that the placebo effect is real and powerful,[1] perhaps even growing stronger,[2] and appreciate more of the associated consequences that relate to expectations and suggestions.[3] And yet, many a skeptic may find some of the ideas described in this book to be pretty far out on a limb. These skeptics won't be convinced until the results have been independently replicated under strictly controlled conditions—nor should they be.

Modern science largely advocates for an overarching approach inspired by reductionism—a preference for lower-level explanations alongside a fundamental confidence in the adequacy of neurobiological mechanisms to explain behavior. This approach holds not only in physics but also in the life sciences and medicine. According to this model, the chemistry of chicken soup more plausibly explains an interaction with a bacterium than would the symbolic blessing from a king's laying on of hands, or the King's Touch—a common practice whereby monarchs of different cultures stroked their subjects with the intent to cure them of various illnesses.[4]

A preference for a bottom-up model to explain our vulnerability and resilience to illness coincides with the marginalization of cultural symbols such as the top-down King's Touch. But a King's Touch can be no less, and possibly more, antibacterial than chicken soup. While the King's Touch is unlikely to act directly on bacteria, it may well influence the immune system of individuals touched by the royal through a cascading process governing their thoughts and emotions in a top-down fashion.[5]

In other words, a special caress from the king may goad into action a higher-order downstream effect, which stems from our social mores, cultural values, and personal expectations. The psychology of faith and belief together with the notions of tradition and culture all seem to play a key role in these suggestive effects. As such, we have seen that people who follow a specific tradition often respond dogmatically to evidence against their tradition; moreover, individuals from different traditions view the same evidence through different lenses. To make things even more complicated, converting from one tradition to another isn't the same as an ordinary shift in belief.[6] So, the psychology of suggestion and the clinical "healing" of suggestion closely relate to the nature and rationality of faith, entrenching us in the core assumptions of our traditions. For those individuals who hold the king in high esteem, therefore, the King's Touch may goad into action special therapeutic physiology.

We already established that our mindset and perceptions are important to our health.[7] But let's take a closer look at two illustrative examples. For example, consider the hotel maid study. Housekeeping entails physical activities, including making beds, cleaning surfaces, and removing garbage; however, seldom would we frame it as exercise. But what if we did?

Two out of three housekeepers reported that they didn't exercise, and more than one-third of those reported they didn't get any exercise at all. Measuring their body fat, waist-to-hip ratio, blood pressure, weight, and body mass index revealed that all these indicators matched the maids' *perceived* amount of exercise, rather than their *actual* activity. But then experimenters suggested to half of the hotel housekeepers enrolled in the study that their work routine was the equivalent of a workout. They informed the maids that their everyday work chores entailed sufficient exercise, enough to exceed the recommendations for an active lifestyle as outlined by the US surgeon general. Through their regular routine of sweeping, wiping, and cleaning, the room attendants were burning the same number of calories as they would if they hopped on a treadmill and sweated it out for an hour in the gym. In line with the findings of the shake study (Chapter 4), the housekeepers experienced substantial health benefits following this simple suggestion. They lost weight, lowered their blood pressure, and minimized their body mass index, without changing their daily routine.[8]

Similarly, researchers showed that we can shape our health and well-being independently of actual physical activity by learning how much we actually move. They gave naïve participants tracking watches to wear for four weeks: one group received a "Booster" watch that inflated their step counts—from seven thousand to an increase of 40 percent, 9,800 steps—whereas a second group wore a "Downer" watch that deflated step counts—lowering them by 40 percent to around 4,200 steps per day. A third and a fourth group received genuine watches that reported true step counts.

Participants in the groups that wore the genuine and Booster watches started eating better, consuming fewer high-fat foods and more vegetables. Their aerobic fitness had risen a bit, although they weren't exercising more—both walked an average of seven thousand steps per day. On the other hand, people wearing the Downer watch displayed slightly lower self-esteem, dimmer moods, poorer eating habits, and small increases in their resting heart rates and blood pressure—all indicators of worsening health, although their true step counts, empirically, were on par with the other groups'. This study shows that suggestions about our exercise habits can change our motivation and goals. Even if those mindsets don't reflect reality, they can change our physiology.[9]

We all know that physical activity is a critical determinant of our health and well-being. As more and more aspects of life become trackable—from fitness, insomnia, posture, and stress levels to screen time, work hours, and productivity—it turns out that not only does our actual physical activity matter, but so does our mindset about it. Specifically, we should all be aware that our health interventions would be more successful if we more deliberately—and more effectively—harness the power of suggestion and mindset.

But how can we use the power of mindset and form our own suggestions to achieve this goal? In other words, how can we know what to believe? And when we finally settle on what to believe, how can we be sure of it? Where should our confidence come from? After all, the ability to possess deep conviction about something that another person finds no reason to believe in poses a formidable challenge not just to our society but also to modern psychologists and brain scientists.[10]

Typically, the degree of confidence that people have in their beliefs, opinions, and mindsets has a very weak connection to the truth. In other words, our profound persuasion about something hardly makes it real or, to put it a different way, scarcely makes us right. The lack of correlation between confidence and truth is a striking psychological fact: when people are adamantly certain that they are right, and don't entertain the possibility that they even *might* be wrong, life becomes difficult. The best thing we can do is to constantly question our beliefs, opinions, and mindsets.

Personally, I make this mistake all the time. I'm pretty sure of my opinions. I usually express them with sureness and am generally convinced that my beliefs are right. But what if I'm wrong? Knowing what I know and have shared with you throughout this book, I ought to give this possibility some weight, ought I not? So, I try. But often, I'm just too honest-to-goodness certain to so much as pause and doubt myself.

The right thing to do is to approach things in the same way we would any charged topic we try to educate ourselves about—say, policing and race in the US or the use of psychedelics for personal growth. First, identify and critically assess the dominant narrative of hot-button issues using evidence-based knowledge drawn from reliable sources—for example, by evaluating information for credibility, bias, and authority. Second, acknowledge the role of social media in spreading polarization and disinformation while keeping an eye out for dog whistles, political speech, and overarching agendas. Third, recognize that any kind of friction on a given topic is a good indication that multiple ideas are entering the fray—that's a good thing. This tension should serve as a type of internal, deliberative dialogue that helps not just with forming a helpful mindset, but also with facilitating a better understanding of and cooperation with situations, ourselves, and others.

Many of the top-down examples peppered throughout this book are noteworthy because they rely on the importance of how our minds construe a situation.[11] For the experiments involving the hotel maids and tracking watches, while these results were *statistically* significant, often such effects are small—unlikely to be *clinically* significant—without follow-up measures, and with findings that don't always easily replicate.[12] But these nascent examples are not demonstrations of exotic unicorns; rather, they herald a new field and a research trajectory that we should follow and pursue with gusto. While I am optimistic that future studies will further elucidate these provocatively curious leads, the proof remains in the pudding.

That's why, regardless of data, replications, or scientific understanding of the underlying mechanisms, we need to reserve a certain measure of skepticism as we move forward. For example, it's possible that the hotel maids might have behaved differently after they had received the information from the researchers and that they were being more active and eating more healthfully, and thereby improved their measurements. However, the researchers said that they had not observed any such changes. Being skeptical is important, but we must also be careful: overzealous skepticism and exclusive reliance on bottom-up mechanisms as a philosophical ordinance likely expose a problem.

In general, many people do not believe that expectations and attitudes can produce this kind of *objective* change in the physical body. The intimation is that they can bring about *subjective* types of findings—say, to our *perception* of pain or our *feeling* of depression—but not something as objective as, say, cutting ten pounds off our waistlines or markedly reducing our blood pressure. This book challenges this old assumption—that mindset can alter only subjective perception. Having read through the examples and stories of the suggestible brain, it should become obvious that the idea that the placebo effect applies only to subjective things is one that we must dismiss and dispel. I am not yet saying that you shouldn't purchase that overpriced Peloton or "Juicing 101" book. Instead, you should remember that your mind is already naturally susceptible to suggestion. Our ability to alter our perceptions and reshape reality is one of the most inexpensive and practical higher brain functions we possess to improve our lives.

Back when we discussed why antidepressants, despite the problematic evidence, are still going strong, we saw how philosopher Kuhn explained the way transformation comes about in science. According to Kuhn, research data are not enough to drive a meaningful change; the old guard needs to phase out before new ideas take hold. But moving the needle on scientific consensus and establishing a groundswell change is a Herculean task. Doing science right—rigorously and systematically while controlling for all the right variables—is expensive, time-consuming, and challenging. One researcher on a limited budget can't do much more than get a conversation going. The good news is that with increasingly more researchers, scientists, practitioners, and clinicians chiming in, a critical mass of voices adds up to a chorus of scholars whose message is more difficult to ignore.

Slowly but surely, the importance of suggestion and mindset is starting to factor into not just clinical and research issues but also our cultural landscape. In medicine, investigators are experimenting with psychedelics to treat everything from autism and opioid addiction to anorexia and the anxieties experienced by the terminally ill.[13] Across society, advocates of psychedelics often highlight their positive potential, and the number of users is steadily increasing. On the other hand, anecdotal reports highlight individuals especially prone to psychedelically facilitated conspiratorial thinking—from David Icke, who popularized a conspiracy theory about shapeshifting reptilian aliens controlling the world and linking 5G mobile technology[14] to COVID-19 symptoms;[15] to the QAnon Shaman, Jake Angeli, a visible player in the storming of the US Capitol on January 6, 2021, and a self-styled psychedelic guru;[16] all the way to Mikki Willis, a documentary film producer peddling an assortment of COVID-19 conspiracies.[17] As such, it's undeniable that the looming psychedelic revolution pivots around the science of suggestion and holds overarching implications to our social and cultural worlds. If you so much as visit Burning

Man, or are an active participant in the nightlife community, or even a guest at the electronic music scene, you will recognize that the recreational use of psychedelics is pervasive and that the change that individuals may go through, to the extent it occurs, is not always deterministically predictable.[18]

The burgeoning field of the suggestible brain is making its first steps. We still don't fully understand how suggestion works; however, we have come a long way in designing experiments that allow us to tease apart the different elements that likely contribute to some of its effects. Because even our best experiments cannot possibly account for all the mediators and moderators that may be responsible for the results we see, we can entertain at least two approaches. One approach could follow the practical sentiment of "if it works, I'll take it"; after all, clinical interventions put more emphasis on getting you better and less on how you got there. Another approach could take a more scientific attitude in which the goal is to understand and map out the fundamental mechanisms that make it all work out. Throughout this book, you have gotten a taste for both.

We can glean meaningful insights from inspiring human stories, such as how an athlete in her nineties can teach us about living longer and the role of the mind therein, for example.[19] But to seed a global transformation, we need to examine a wide array of converging evidence, well-designed experiments, underlying mechanisms, and scientific data. Here we have learned about how top-down suggestions can affect our brain states and behavior.[20] Even with imperfect memory, medical challenge tests, and ethical restrictions imposed on research studies, we are making strides toward obtaining a more scientific understanding of how suggestion and suggestibility affect humans. The King's Touch is not a "fun fact" for kings in the same way that Machiavelli's *The Prince* is not just for princes.[21] Against a backdrop of open-label placebos, sham surgeries, antidepressants, psychedelics, and social contagion, our suggestible brains find discernible patterns in patternless, random noise,[22] hear voices when no sound is present,[23] and change our physiology as a function of our thoughts.[24]

Moreover, suggestions bind the "science"—the cognitive, rational, and logical aspects of our world—with the "magical"—the emotional, uncertain, and metaphysical experiences that make up life. In this regard, suggestions entail an evolutionary advantage: they permit us to learn quickly, bond with social groups, and explore a unique dimension of the human experience—spirituality and belief.[25] You have gotten a peek into the top-down science of suggestion; if you learned anything from this book, let it be that suggestions will move you whether or not you notice or believe them.

By learning about the art and science of suggestion, you can empower and protect yourself, change your reality, boost your immunity, alter your physiology, help others, and create social impact by making the world a better place. Civilized deception, not to be confused with nasty subterfuge, plays an important role in our normal lives, and we shouldn't shy away from studying and applying such gray-area phenomena using both the magical arts and the relevant sciences.[26] To be sure, suggestion and suggestibility form key concepts in this quest.[27]

Suggestion joins faith, belief, and purpose—mighty forces that are hard to measure with bottom-up science. "Not everything that can be counted counts, and not everything that counts can be counted" is a quote attributed to Albert Einstein. As a physicist, Einstein was a natural reductionist. He believed in bottom-up theories that have already earned their important place in one of the most sublime intellectual triumphs of the only non-extinct Homo genus—science. But he also acknowledged that a bottom-up approach may be insufficient to capture all of nature. Having read this book, you should feel comfortable with the notion that not everything that can be measured matters, and not everything that matters can be measured. As we get deeper into a newer, interdisciplinary, top-down science of suggestion, perhaps it's a good idea to keep Einstein's adage in mind.

ACKNOWLEDGMENTS

I gratefully acknowledge support from my family members, friends, mentors, colleagues, and many students over the years and decades and across multiple research environments. Special thanks to Nicholas Isy Barrett for assistance with all the graphic art.

REFERENCES

Introduction

1. Raz, A., and C. S. Harris. *Placebo Talks: Modern Perspectives on Placebos in Society.* Oxford: Oxford University Press, 2016.

2. Blakeslee, S. "Placebos Prove So Powerful Even Experts Are Surprised; New Studies Explore the Brain's Triumph Over Reality." *New York Times,* October 1998; Ikemi, Y. "A Psychosomatic Study of Contagious Dermatitis." *Kyushu Journal of Medical Science* (1962).

3. Kirsch, I., and G. Sapirstein. "Listening to Prozac but Hearing Placebo: A Meta-Analysis of Antidepressant Medication." *Prevention & Treatment* 1, (June 1998).

4. Webster, S. T. "Mind over Milkshakes: How Your Mindset Guides Physical Response to Food." *IDEA Health & Fitness* (June 2014): 114.

5. Park, C., F. Pagnini, and E. Langer. "Glucose Metabolism Responds to Perceived Sugar Intake More Than Actual Sugar Intake." *Scientific Reports* 10, no. 1 (September 2020). https://doi .org:10.1038/s41598-020-72501-w

6. Daniel, M. B., et al. "The False Memory Diet: False Memories Alter Food Preferences." *Handbook of Behavior, Food and Nutrition* (January 2011); Willett, W., et al. "Food in the Anthropocene: The EAT–*Lancet* Commission on Healthy Diets from Sustainable Food Systems." *The Lancet Commissions* 393, no. 10170 (February 2019). https://doi.org:10.1016/S0140-6736(18)31788-4

7. Crum, A. J., and E. J. Langer. "Mind-Set Matters: Exercise and the Placebo Effect." *Psychological Science* 18, no. 2 (2007): 165–171.

8. Zahrt, O. H., et al. "Effects of Wearable Fitness Trackers and Activity Adequacy Mindsets on Affect, Behavior, and Health: Longitudinal Randomized Controlled Trial." *Journal of Medical Internet Research* 25, e40529 (2023). https://doi.org:10.2196/40529

9. Langer, E., M. Djikic, M. Pirson, A. Madenci, and R. Donohue. "Believing Is Seeing: Using Mindlessness (Mindfully) to Improve Visual Acuity." *Psychological Science* 21, no. 5 (May 2010): 661; Raz, A., Z. R. Zephrani, H. R. Schweizer, and G. P. Marinoff. "Critique of Claims of Improved Visual Acuity After Hypnotic Suggestion." *Optometry and Vision Science* 81, no. 11 (November 2004): 872–879; Raz, A., G. P. Marinoff, Z. R. Zephrani, H. R. Schweizer, and M. I. Posner. "See Clearly: Suggestion, Hypnosis, Attention, and Visual Acuity." *International Journal of Clinical and Experimental Hypnosis* 52, no. 2 (April 2004): 159–187. https://doi.org:10.1076 /iceh.52.2.159.28097; Raz, A., et al. "Substrates of Negative Accommodation." *Binocular Vision & Strabismus* 19, no. 2 (2004): 71–74.

10. Lichtenbelt, W. v. M. "Who Is the Iceman?" *Temperature* 4, no. 3 (June 2017): 202–205. https:// doi.org:https://doi.org/10.1080/23328940.2017.1329001

11.   Palsson, O. S., and S. Ballou. "Hypnosis and Cognitive Behavioral Therapies for the Management of Gastrointestinal Disorders." *Current Gastroenterology Reports* 22, no. 7 (June 2020). https://doi .org:https://doi.org/10.1007/s11894-020-00769-z

12.   Jensen, M. P., et al. "Effects of Hypnosis, Cognitive Therapy, Hypnotic Cognitive Therapy, and Pain Education in Adults with Chronic Pain: A Randomized Clinical Trial." *Pain* 161, no. 10 (2020): 2284–2298.

Chapter 1

1.   Blakeslee, S. "This Is Your Brain Under Hypnosis." *New York Times*, November 2005.

2.   Dienes, Z., et al. "Hypnotic Suggestibility, Cognitive Inhibition, and Dissociation." *Consciousness and Cognition* 18, no. 4 (December 2009): 837–847. https://doi.org:10.1016/j.concog .2009.07.009

3.   Landry, M., M. Lifshitz, and A. Raz. "Brain Correlates of Hypnosis: A Systematic Review and Meta-analytic Exploration." *Neuroscience and Biobehavioral Reviews* 81, Pt. A (October 2017): 75–98. https://doi.org:10.1016/j.neubiorev.2017.02.020

4.   Taru, K., H. S. Zamansky, and M. L. Block. "Is the Hypnotized Subject Lying?" *Journal of Abnormal Psychology* 103, no. 2 (May 1994): 184–191. https://doi.org:10.1037/0021-843X.103.2.184; Raz, A., T. Shapiro, J. Fan, and M. I. Posner. "Hypnotic Suggestion and the Modulation of Stroop Interference." *Archives of General Psychiatry* 59, no. 12 (December 2002): 1155–1161; Oakley, D. A., in *The Oxford Handbook of Hypnosis: Theory, Research, and Practice*, edited by M. R. Nash and A. J. Barnier, 365–392. Oxford: Oxford University Press, 2008; Kirsch, I., C. E. Silva, J. E. Carone, J. D. Johnston, and B. Simon. "The Surreptitious Observation Design: An Experimental Paradigm for Distinguishing Artifact from Essence in Hypnosis." *Journal of Abnormal Psychology* 98, no. 2 (May 1989): 132–136; Raz, A., J. Fan, and M. I. Posner. "Hypnotic Suggestion Reduces Conflict in the Human Brain." *Proceedings of the National Academy of Sciences of the United States of America* 102, no. 28 (June 2005): 9978–9983; Raz, A., I. Kirsch, J. Pollard, and Y. Nitkin-Kaner. "Suggestion Reduces the Stroop Effect." *Psychological Science* 17, no. 2 (February 2006): 91–95.

5.   Spanos, N. P. *Multiple Identities and False Memories: A Sociocognitive Perspective*. American Psychological Association, 1996; Spanos, N. P. "Multiple Identity Enactments and Multiple Personality Disorder: A Sociocognitive Perspective." *Psychological Bulletin* 116, no. 1 (July 1994): 143–165.

6.   Gheorghiu, V. A. "The Development of Research on Suggestibility: Critical Considerations." Conference paper. Springer Berlin, Heidelberg, 2011.

7.   Gauld, A. *A History of Hypnotism*. Cambridge University Press, 1993.

8.   Raz, A. "Hypnosis: A Twilight Zone of the Top-Down Variety: Few Have Never Heard of Hypnosis but Most Know Little About the Potential of This Mind-Body Regulation Technique for Advancing Science." *Trends in Cognitive Sciences* 15, no. 12 (December 2011): 555–557. https:// doi.org:10.1016/j.tics.2011.10.002

9.   Shor, R. E., and E. C. Orne. "Harvard Group Scale of Hypnotic Susceptibility, Form A." 1962.

10.   Weitzenhoffer, A. M. *Stanford Hypnotic Susceptibility Scale: Forms A and B, for Use in Research Investigations in the Field of Hypnotic Phenomena*. Consulting Psychologists Press, 1959.

11.   Raz, A. "Suggestibility and Hypnotizability: Mind the Gap." *American Journal of Clinical Hypnosis* 49, no. 3 (January 2007): 205–210. https://doi.org:https://doi.org/10.1080/00029157.2007.10401582

12.   Kirsch, I. "Suggestibility or Hypnosis: What Do Our Scales Really Measure?" *International Journal of Clinical and Experimental Hypnosis* 45, no. 3 (July 1997): 212–225.

13.   Raz, A. "Suggestibility and Hypnotizability: Mind the Gap." *American Journal of Clinical Hypnosis* 49, no. 3 (January 2007): 205–210. https://doi.org/10.1080/00029157.2007.10401582; Lifshitz, M., et al. "On Suggestibility and Placebo: A Follow-Up Study." *American Journal of Clinical Hyp-*

REFERENCES

nosis 59, no. 4 (April 2017): 385–392; Sheiner, E. O., M. Lifshitz, and A. Raz. "Placebo Response Correlates with Hypnotic Suggestibility." *Psychology of Consciousness: Theory, Research, and Practice* 3, no. 2 (2016): 146–153. https://doi.org:10.1037/cns0000074

14. Hartogsohn, I. "Set and Setting, Psychedelics and the Placebo Response: An Extra-pharmacological Perspective on Psychopharmacology." *Journal of Psychopharmacology* 30, no. 12 (December 2016): 1259–1267.

15. Olson, J. A., L. Suissa-Rocheleau, M. Lifshitz, A. Raz, and S. P. L. Veissière. "Tripping on Nothing: Placebo Psychedelics and Contextual Factors." *Psychopharmacology* 237, no. 5 (May 2020): 1371–1382. https://doi.org:10.1007/s00213-020-05464-5

16. Abraham, H. H. L. "The Suggestible Personality: A Psychological Investigation of Susceptibility to Persuasion." *Acta Psychologica* 20, (1962): 167–184. https://doi.org:10.1016/0001-6918(62)90016-1

17. Kohen, D. P., and K. Olness. *Hypnosis and Hypnotherapy with Children*. 4th ed. Routledge, 2011; "The Suggestibility of Children: Evaluation by Social Scientists." Excerpt from Bruck, M. and S. J. Ceci. "Amicus Brief for the Case of *State of New Jersey v. Michaels* Presented by Committee of Concerned Social Scientists." 1994. Famous Trials, edited by D. O. Linder. University of Missouri–Kansas City School of Law. https://famous-trials.com/mcmartin/908-suggesta bility; Anbar, R. D. *Changing Children's Lives with Hypnosis: A Journey to the Center*. Rowman & Littlefield, 2021.

18. Ray, W. J., and D. M. Tucker. "Evolutionary Approaches to Understanding the Hypnotic Experience." *International Journal of Clinical and Experimental Hypnosis* 51, no. 3 (July 2003): 256–281. https://doi.org:https://doi.org/10.1076/iceh.51.3.256.15520; Ginsburg, S., and E. Jablonka. "Evolutionary Transitions in Learning and Cognition." *Philosophical Transactions of the Royal Society B* 376, no. 1821 (March 2021): 20190766. https://doi.org/10.1098/rstb.2019.0766; Jamieson, G. *Hypnosis and Conscious States: The Cognitive Neuroscience Perspective*. Oxford: Oxford University Press, 2007.

19. Morgan, A. H., and E. R. Hilgard. "Age Differences in Susceptibility to Hypnosis." *International Journal of Clinical and Experimental Hypnosis* 21, no. 2 (1973): 78–85. https://doi.org:10.1080/00207147308409308; London, P., and L. M. Cooper. "Norms of Hypnotic Susceptibility in Children." *Developmental Psychology* 1, no. 2 (1969): 113–124. https://doi.org/10.1037/h0027002; Linden, J. H. "Relationship Factors in the Theater of the Imagination: Hypnosis with Children and Adolescents." *American Journal of Clinical Hypnosis* 62, no. 1–2 (July 2019): 60–73.

20. Piccione, C., E. R. Hilgard, and P. G. Zimbardo. "On the Degree of Stability of Measured Hypnotizability Over a 25-Year Period." *Journal of Personality and Social Psychology* 56, no. 2 (February 1989): 289–295.

21. Spiegel, H. "The Grade 5 Syndrome: The Highly Hypnotizable Person." *International Journal of Clinical and Experimental Hypnosis* 22, no. 4 (October 1974): 303–319. https://doi.org/10.1080/00207147408413010; Orne, M. T. "The Nature of Hypnosis: Artifact and Essence." *Journal of Abnormal Social Psychology* 58, no. 3 (1959): 277–299.

22. Raz, A. "Suggestibility and Hypnotizability: Mind the Gap." *American Journal of Clinical Hypnosis* 49, no. 3 (January 2007): 205–210. https://doi.org/10.1080/00029157.2007.10401582; Lifshitz, M., E. O. Sheiner, J. A. Olson, R. Theriault, and A. Raz. "On Suggestibility and Placebo: A Follow-Up Study." *American Journal of Clinical Hypnosis* 59, no. 4 (April 2017): 385–392. https://doi.org/10.1080/00029157.2016.1225252

23. Friston, K. "Does Predictive Coding Have a Future?" *Nature Neuroscience* 21 (2018): 1019–1021. https://doi.org/10.1038/s41593-018-0200-7; Kihlstrom, J. F. "Neuro-Hypnotism: Prospects for Hypnosis and Neuroscience." *Cortex* 49, no. 2 (February 2013): 365–374. https://doi.org/10.1016/j.cortex.2012.05.016; Huber, A., F. Lui, D. Duzzi, G. Pagnoni, and C. Adolfo-Porro. "Structural and Functional Cerebral Correlates of Hypnotic Suggestibility." *PLOS One* 9, no. 3 (March 2014). https://doi.org/10.1371/journal.pone.0093187

24. Parr, T., G. Pezzulo, and K. J. Friston. *Active Inference: The Free Energy Principle in Mind, Brain, and Behavior.* MIT Press, 2022; Hohwy, J. *The Predictive Mind.* Oxford: Oxford University Press, 2014; Friston, K. "Waves of Prediction." *PLOS Biology* 17, no. 10 (October 2019). https://doi.org/10.1371/journal.pbio.3000426

25. Cepelewicz, J. "To Make Sense of the Present, Brains May Predict the Future." *Quanta,* July 10, 2018.

26. Poulsen, B. C. "Correlates of Imaginative Suggestibility and Hypnotizability in Children." Doctoral diss., University of Massachusetts Amherst, 2000; Kelman, H. C. "Effects of Success and Failure on 'Suggestibility' in the Autokinetic Situation." *The Journal of Abnormal and Social Psychology* 45, no. 2 (April 1950): 267–285. https://doi.org/10.1037/h0062561; Abazari, N., L. Heydarinasab, H. Yaghubi, and H. Farahani. "Predictability of Pain Intensity and Psychological Distress by Suggestibility and Attitude to Menstruation Among Female University Students." *Research Square* (July 2020). https://doi.org/10.21203/rs.3.rs-42450/v1

27. Hilgard, E. R., in *Human Suggestibility: Advances in Theory, Research, and Application,* edited by J. F. Schumaker, 37–58. Taylor & Frances/Routledge, 1991.

28. Braaten, E. B., and D. Norman. "Intelligence (IQ) Testing." *Pediatric Review* 27, no. 11 (November 2006): 403–408.

29. Mackintosh, N. J. *IQ and Human Intelligence.* 2nd ed. Oxford: Oxford University Press, 2011.

30. Gottfredson, L. S. "The General Intelligence Factor." *Scientific American Presents* 9, no. 4 (1998): 24–29.

31. Canivez, G. L. "Psychometric Versus Actuarial Interpretation of Intelligence and Related Aptitude Batteries." In *The Oxford Handbook of Child Psychological Assessment,* edited by D. H. Saklofske, C. R. Reynolds, and V. L. Schwean, 84–112. Oxford: Oxford University Press, 2013.

32. Kamphaus, R. W., A. P. Winsoe, E. W. Rowe, and S. Kim in *Contemporary Intellectual Assessment: Theories, Tests, and Issues.,* 4th ed., edited by D. P. Flanagan and E. M. McDonough, 56–70. The Guilford Press, 2018.

33. Rust, J., and S. Golombok. *Modern Psychometrics: The Science of Psychological Assessment.* 3rd ed. Routledge/Taylor & Francis Group, 2009.

34. Council, J. R., in *Clinical Hypnosis and Self-Regulation: Cognitive-Behavioral Perspectives.* Dissociation, Trauma, Memory, and Hypnosis Book Series, edited by I. Kirsch, A. Capafons, E. Cardeña-Buelna, and S. Amigó, 119–140. American Psychological Association, 1999.

35. Abraham, H. H. L. "The Suggestible Personality: A Psychological Investigation of Susceptibility to Persuasion." *Acta Psychologica* 20, (1962): 167–184. https://doi.org:10.1016/0001-6918(62)90016-1

36. Weitzenhoffer, A. M. *Stanford Hypnotic Susceptibility Scale: Forms A and B, for Use in Research Investigations in the Field of Hypnotic Phenomena.* Consulting Psychologists Press, 1959.

37. Laurence, J. R., D. Beaulieu-Prévost, and T. Du Chéné. *Measuring and Understanding Individual Differences in Hypnotizability.* Oxford: Oxford University Press, 2008; Hilgard, E. R. *Hypnotic Susceptibility.* Harcourt, Brace & World, 1965.

38. Switras, J. E. "A Comparison of the Eye-Roll Test for Hypnotizability and the Stanford Hypnotic Susceptibility Scale: Form A." *American Journal of Clinical Hypnosis* 17, no. 1 (1974): 54–55. https://doi.org:10.1080/00029157.1974.10403707.

39. Spiegel, H., and D. Spiegel. *Trance and Treatment: Clinical Uses of Hypnosis.* Basic Books, 1978.

40. Spiegel, H. "An Eye-Roll Test for Hypnotizability." *American Journal of Clinical Hypnosis* 53, no. 1 (July 2010): 15–18.

41. Spiegel, H., and D. Spiegel. *Trance and Treatment: Clinical Uses of Hypnosis.* Basic Books, 1978.

42. Hilgard, E. R. "Illusion That the Eye-Roll Sign Is Related to Hypnotizability." *Archives of General Psychiatry* 39, no. 8 (August 1982): 963–966.

43. Switras, J. E. "A Comparison of the Eye-Roll Test for Hypnotizability and the Stanford Hypnotic Susceptibility Scale: Form A." *American Journal of Clinical Hypnosis* 17, no. 1 (1974): 54–55. https://doi.org:10.1080/00029157.1974.10403707; Perry, C., R. Nadon, and J. Button in *Contemporary Hypnosis Research*, edited by E. Fromm and M. R. Nash, 459–490. Guilford Press, 1992; Sheehan, P. W., and K. M. McConkey. *Hypnosis and Experience: The Exploration of Phenomena and Process.* Erlbaum, 1982.

44. Spiegel, H. "An Eye-Roll Test for Hypnotizability." *American Journal of Clinical Hypnosis* 53, no. 1 (July 2010): 15–18.

45. Piccione, C., E. R. Hilgard, and P. G. Zimbardo. "On the Degree of Stability of Measured Hypnotizability Over a 25-Year Period." *Journal of Personality and Social Psychology* 56, no. 2 (February 1989): 289–295.

46. Raz, A., T. Hines, J. Fossella, and D. Castro. "Paranormal Experience and the COMT Dopaminergic Gene: A Preliminary Attempt to Associate Phenotype with Genotype Using an Underlying Brain Theory." *Cortex* 44, no. 10 (November–December 2008): 1336–1341. https://doi.org:10.1016/j.cortex.2007.07.011

47. Kirsch, I., A. Capafons, E. Cardeña-Buelna, and S. Amigó. *Clinical Hypnosis and Self-Regulation: Cognitive-Behavioral Perspectives.* American Psychological Association, 1999.

48. Lorenz, M. P., J. R. Ramsey, J. M. Andzulis, and G. R. Franke. "The Dark Side of Cultural Intelligence: Exploring Its Impact on Opportunism, Ethical Relativism, and Customer Relationship Performance." *Business Ethics Quarterly* 30, no. 4 (April 2020): 552–590. https://doi.org:10.1017/beq.2020.2; Cardeña, E., and S. Krippner. "The Cultural Context of Hypnosis." In *Handbook of Clinical Hypnosis*, edited by S. J. Lynn, J. W. Rhue, and I. Kirsch, 743–771. American Psychological Association, 2010.

49. Shigeharu, M. "The Present State of Hypnosis in Japan." *The Japanese Journal of Educational & Social Psychology* 7, no. 1 (1967): 65–71. https://doi.org:10.11558/jjesp1960.7.65

50. Shimizu, T., and M. Kodoma. "The Relation Between Notions of Hypnotic States and Attitude Towards Hypnosis." *Tsubaka Psychological Research* 23, (2001): 219–227; Koizumi, S. "Investigation into University Students in Distress at Japanese Society." *Journal of Hypnosis* (2001): 40–46.

51. Champigny, C. M., and A. Raz. "Transcultural Factors in Hypnotizability Scales: Limits and Prospects." *American Journal of Clinical Hypnosis* 58, no. 2 (October 2015): 171–194.

52. Weitzenhoffer, A. M. *Hypnotism: An Objective Study in Suggestibility.* Wiley, 1953.

53. Trouton, D. S. "Placebos and Their Psychological Effects." *Journal of Mental Science* 103 (1957): 344–354.

54. Review of *Personality and Motivation Structure and Measurement*, by R. B. Cattell. *Postgrad Medical Journal* 34, no. 393 (July 1958): 398.

55. Evans, F. J. "The Independence of Suggestibility, Placebo Response, and Hypnotizability." Conference paper, 145–154. Springer Berlin, Heidelberg, 2011.

56. Lifshitz, M., E. O. Sheiner, J. A. Olson, R. Theriault, and A. Raz. "On Suggestibility and Placebo: A Follow-Up Study." *American Journal of Clinical Hypnosis* 59, no. 4 (April 2017): 385–392. https://doi.org/10.1080/00029157.2016.1225252; Sheiner, E. O., M. Lifshitz, and A. Raz. "Placebo Response Correlates with Hypnotic Suggestibility." *Psychology of Consciousness: Theory, Research, and Practice* 3, no. 2 (2016): 146–153. https://doi.org:10.1037/cns0000074; Olson, J. A., L. Suissa-Rocheleau, M. Lifshitz, A. Raz, and S. P. L. Veissière. "Tripping on Nothing: Placebo Psychedelics and Contextual Factors." *Psychopharmacology* 237, no. 5 (May 2020): 1371–1382. https://doi.org:10.1007/s00213-020-05464-5; Raz, A. "Hypnobo: Perspectives on Hypnosis and Placebo." *American Journal of Clinical Hypnosis* 50, no. 1 (July 2007): 29–36; Evers, A. W. M.,

et al. "Implications of Placebo and Nocebo Effects for Clinical Practice: Expert Consensus." *Psychotherapy and Psychosomatics* 87, no. 4 (2018): 204–210. https://doi.org:10.1159/000490354

57. Wagstaff, G. F., in *Human Suggestibility: Advances in Theory, Research, and Application*, edited by J. F. Schumaker, 132–145. Taylor & Frances/Routledge, 1991.

58. Gudjonsson, G. H., and N. K. Clark. "Suggestibility in Police Interrogation: A Social Psychological Model." *Social Behaviour* 1, no. 2 (1986): 83–104.

59. Andriks, J. L., E. F. Loftus, and P. A. Powers. "Eyewitness Accounts of Females and Males." *Journal of Applied Psychology* 64, no. 3 (1979): 339–347.

60. Zhao, J., et al. "Association Between Daily Alcohol Intake and Risk of All-Cause Mortality: A Systematic Review and Meta-Analyses." *JAMA Network Open* 6, no. 3 (2023): e236185. https://doi.org:10.1001/jamanetworkopen.2023.6185

61. Cordova, A. C., and B. E. Sumpio. "Polyphenols Are Medicine: Is It Time to Prescribe Red Wine for Our Patients?" *International Journal of Angiology* 18, no. 3 (2009): 111–117. https://doi.org:10.1055/s-0031-1278336

62. Rabin, R. C. "Moderate Drinking Has No Health Benefits, Analysis of Decades of Research Finds." *New York Times*, April 4, 2023.

63. Almenberg, J., and A. Dreber. "When Does the Price Affect the Taste? Results from a Wine Experiment." SSE/EFI Working Paper Series in Economics and Finance No. 717, Stockholm School of Economics, The Economic Research Institute, Stockholm, 2009.

64. Morrot, G., F. Brochet, and D. Dubourdieu. "The Color of Odors." *Brain Lang* 79, no. 2 (November 2001): 309–320. https://doi.org:10.1006/brln.2001.2493

65. Cicchetti, D. "Assessing the Reliability of Blind Wine Tasting: Differentiating Levels of Clinical and Statistical Meaningfulness." *Journal of Wine Economics* 2, no. 2 (2007): 196–202.

66. Morrot, G., F. Brochet, and D. Dubourdieu. "The Color of Odors." *Brain Lang* 79, no. 2 (November 2001): 309–320. https://doi.org:10.1006/brln.2001.2493

67. Morrot, G., F. Brochet, and D. Dubourdieu. "The Color of Odors." *Brain Lang* 79, no. 2 (November 2001): 309–320. https://doi.org:10.1006/brln.2001.2493

68. Marlatt, G. A., and D. J. Rohsenow. "THE THINK-DRINK EFFECT." *Psychology Today*, December 1981.

69. Cook, P. J. *Paying the Tab: The Economics of Alcohol Policy*. Princeton University Press, 2007.

70. Blanchette, J. G., F. J. Chaloupka, and T. S. Naimi. "The Composition and Magnitude of Alcohol Taxes in States: Do They Cover Alcohol-Related Costs?" *Journal of Studies on Alcohol and Drugs* 80, no. 4 (2019): 408–414. https://doi.org:10.15288/jsad.2019.80.408

71. Semple, K. W., and A. Westbrook. "A Toast to Raising Alcohol Taxes." *New York Times*, May 4, 2023.

72. Kihlstrom, J. F. "Placebo: Feeling Better, Getting Better, and the Problems of Mind and Body." *McGill Journal of Medicine* 11, no. 2 (November 2008): 212–214.

73. Pfaff, D. W., and S. Sherman. *Laws of Human Behavior: Steps Toward Hard Science*. MIT Press, 2024.

Chapter 2

1. Henneberg, M., and B. J. George. "Possible Secular Trend in the Incidence of an Anatomical Variant: Median Artery of the Forearm." *American Journal of Physical Anthropology* 96, no. 4 (April 1995): 329–334. https://doi.org:10.1002/ajpa.1330960402

2. Lucas, T., J. Kumaratilake, and M. Henneberg. "Recently Increased Prevalence of the Human Median Artery of the Forearm: A Microevolutionary Change." *Journal of Anatomy* 237, no. 4 (October 2020): 623–631. https://doi.org:10.1111/joa.13224

3. Leerssen, J. "Culture, Humanities, Evolution: The Complexity of Meaning-Making over Time." *Philosophical Transactions of the Royal Society of London B* 376 (2021): 20200043. https://doi .org:10.1098/rstb.2020.0043

4. Wells, D. A. "Plasticity-Led Evolution and Human Culture." *Integrative Psychological and Behavioral Science* 55, no. 4 (2021). https://doi.org:10.1007/s12124-021-09607-x

5. German, A., G. Mesch, and Z. Hochberg. "People Are Taller in Countries with Better Environmental Conditions." *Frontiers in Endocrinology* 11, (March 2020): 106. https://doi.org:10.3389 /fendo.2020.00106

6. Creanza, N., O. Kolodny, and M. W. Feldman. "Cultural Evolutionary Theory: How Culture Evolves and Why It Matters." *Proceedings of the National Academy of Sciences* 114, no. 30 (July 2017): 7782–7789. https://doi.org:doi:10.1073/pnas.1620732114

7. Wilsson, L. *My Beaver Colony*. 1st ed. Doubleday, 1968.

8. Granqvist, P. "Attachment, Culture, and Gene-Culture Co-Evolution: Expanding the Evolutionary Toolbox of Attachment Theory." *Attachment and Human Development* 23, no. 1 (2021): 90–113. https://doi.org:10.1080/14616734.2019.1709086

9. Kobayashi, Y., J. Y. Wakano, and H. Ohtsuki. "Evolution of Cumulative Culture for Niche Construction." *Journal of Theoretical Biology* 472, (2019): 67–76. https://doi.org:10.1016/j .jtbi.2019.04.013

10. "*Eager: The Surprising, Secret Life of Beavers and Why They Matter.*" *Kirkus Reviews* 86, (July 2018); "*Eager: The Surprising, Secret Life of Beavers and Why They Matter.*" *Publishers Weekly* 265, (2018): 44–47; "*EAGER: The Surprising, Secret Life of Beavers and Why They Matter.*" *AudioFile* 27, (2018): 49; Bent, N. "*Eager: The Surprising, Secret Life of Beavers and Why They Matter.*" *Booklist* 114, (2018): 8; Hartle, D. "*Eager: The Surprising, Secret Life of Beavers and Why They Matter.*" *Library Journal* 143, (2018): 112; Holley, P. S. "*Eager: The Surprising, Secret Life of Beavers and Why They Matter.*" *Booklist* 115, (2018): 69; Lovejoy, D. A. "*Eager: The Surprising, Secret Life of Beavers and Why They Matter.*" *Choice: Current Reviews for Academic Libraries* 56, (2019): 637–638.

11. Wilsson, L. *My Beaver Colony*. 1st ed. Doubleday, 1968.

12. Fassi, L., S. Hochman, Z. J. Daskalakis, D. M. Blumberger, and R. C. Kadosh. "The Importance of Individual Beliefs in Assessing Treatment Efficacy: Insights from Neurostimulation Studies." eLife 12:RP88889, (2023). https://doi.org/10.7554/eLife.88889.2

Chapter 3

1. Bazzano, L. A., J. Durant, and P. R. Brantley. "A Modern History of Informed Consent and the Role of Key Information." *Ochsner Journal* 21, no. 1 (2021): 81–85. https://doi.org:10.31486 /toj.19.0105

2. Trials of War Criminals before the Nuremberg Military Tribunals.

3. Kean, S. *The Icepick Surgeon: Murder, Fraud, Sabotage, Piracy, and Other Dastardly Deeds Perpetrated in the Name of Science*. Little, Brown and Company, 2021.

4. Brandt, A. M. "Racism and Research: The Case of the Tuskegee Syphilis Study." *Hastings Center Report* 8, no. 6 (December 1978): 21–29; Alsan, M., and M. Wanamaker. "Tuskegee and the Health of Black Men." *Quarterly Journal of Economics* 133, no. 1 (February 2018): 407–455. https://doi.org:10.1093/qje/qjx029

5. Elliott, D. "In Tuskegee, Painful History Shadows Efforts to Vaccinate African Americans." National Public Radio, February 16, 2021. https://www.npr.org/2021/02/16/967011614/in -tuskegee-painful-history-shadows-efforts-to-vaccinate-african-americans.

6. Faden, R. R., and T. L. Beauchamp. *A History and Theory of Informed Consent*. Oxford: Oxford University Press, 1986.

REFERENCES

7. Milgram, S. *Obedience to Authority: An Experimental View*. HarperCollins Publishers, 2009.

8. Hunt, M. "Research Through Deception." *New York Times Magazine*, September 1982; Bhattacherjee, A. "Social Science Research: Principles, Methods, and Practices." *Textbooks Collection* (2012).

9. Bernstein, D. M., et al. "The False Memory Diet: False Memories Alter Food Preferences." *Handbook of Behavior, Food and Nutrition* (January 2011); Willett, W., et al. "Food in the Anthropocene: The EAT–*Lancet* Commission on Healthy Diets from Sustainable Food Systems." *The Lancet Commissions* 393, no. 10170 (February 2019). https://doi.org:10.1016/S0140-6736(18)31788-4

10. Hunt, M. "Research Through Deception." *New York Times Magazine*, September 1982.

11. Pirandello, L. *Six Characters in Search of an Author and Other Plays*. Penguin Books, 1995.

12. Asch, S. E. "Opinions and Social Pressure." *Scientific American* 193, no. 5 (November 1955): 31–35.

13. Willett, W., et al. "Food in the Anthropocene: The EAT–*Lancet* Commission on Healthy Diets from Sustainable Food Systems." *The Lancet Commissions* 393, no. 10170 (February 2019). https://doi.org:10.1016/S0140-6736(18)31788-4

14. Hunt, M. "Research Through Deception." *New York Times Magazine*, September 1982; Kim, W. O. "Institutional Review Board (IRB) and Ethical Issues in Clinical Research." *Korean Journal of Anesthesiology* 62, no. 1 (January 2012): 3–12. https://doi.org:10.4097/kjae.2012.62.1.3

15. Edgar, H., and D. J. Rothman. "The Institutional Review Board and Beyond: Future Challenges to the Ethics of Human Experimentation." *Milbank Quarterly* 73, no. 4 (1995): 489–506.

16. Boynton, M. H., D. B. Portnoy, and B. T. Johnson. "Exploring the Ethics and Psychological Impact of Deception in Psychological Research." *IRB: Ethics & Human Research* 35, no. 2 (2013): 7–13.

17. Strandberg, T., J. A. Olson, L. Hall, A. Woods, and P. Johansson. "Depolarizing American Voters: Democrats and Republicans Are Equally Susceptible to False Attitude Feedback." *PLOS One* 15, no. 2 (2020).

18. Milgram, S. *Obedience to Authority: An Experimental View*. Harper & Row, 1974; Brannigan, A., I. Nicholson, and F. Cherry. Introduction to "Unplugging the Milgram Machine." Special issue, *Theory & Psychology* 25, no. 5 (October 2015).

19. Foddy, B. "Justifying Deceptive Placebos." In *Placebo Talks: Modern Perspectives on Placebos in Society*, edited by A. Raz and C. Harris, 62–68. Oxford: Oxford University Press, 2016.

20. Kelley, J. M., T. J. Kaptchuk, C. Cusin, S. Lipkin, and M. Fava. "Open-Label Placebo for Major Depressive Disorder: A Pilot Randomized Controlled Trial." *Psychotherapy and Psychosomatics* 81, no. 5 (August 2012): 312–314. https://doi.org:10.1159/000337053; Carvalho, C., et al. "Open-Label Placebo Treatment in Chronic Low Back Pain: A Randomized Controlled Trial." *Pain* 157, no. 12 (December 2016): 2766–2772. https://doi.org:10.1097/j.pain.0000000000000700; Schaefer, M., T. Sahin, and B. Berstecher. "Why Do Open-Label Placebos Work? A Randomized Controlled Trial of an Open-Label Placebo Induction with and Without Extended Information About the Placebo Effect in Allergic Rhinitis." *PLOS One* no. 3 (March 2018); Lembo, A., et al. "Open-Label Placebo vs Double-Blind Placebo for Irritable Bowel Syndrome: A Randomized Clinical Trial." *Pain* 162, no. 9 (September 2021): 2428–2435. https://doi.org:10.1097/j.pain.0000000000002234

21. Blakeslee, S. "Placebos Prove So Powerful Even Experts Are Surprised; New Studies Explore the Brain's Triumph over Reality." *New York Times Magazine*, October 1998; Klopfer, B. "Psychological Variables in Human Cancer." *Journal of Projective Techniques* 21, no. 4 (December 1957): 331–340.

22. Koranyi, N., and K. Rothermund. "When the Grass on the Other Side of the Fence Doesn't Matter: Reciprocal Romantic Interest Neutralizes Attentional Bias Towards Attractive Alternatives."

Journal of Experimental Social Psychology 48, no. 1 (2012): 186–191. https://doi.org:https://doi.org/10.1016/j.jesp.2011.06.012

23. Johansson, P., L. Hall, S. Sikstrom, and A. Olsson. "Failure to Detect Mismatches Between Intention and Outcome in a Simple Decision Task." *Science* 310, no. 5745 (October 2005): 116–119. https://doi.org:10.1126/science.1111709

24. Kuhn, G., J. A. Olson, and A. Raz. "Editorial: The Psychology of Magic and the Magic of Psychology." *Frontiers in Psychology* 7, no. 1358 (2016). https://doi.org:10.3389/fpsyg.2016.01358

25. Raz, A., T. Shapiro, J. Fan, and M. I. Posner. "Hypnotic Suggestion and the Modulation of Stroop Interference." *Archives of General Psychiatry* 59, no. 12 (December 2002): 1155–1161. https://pubmed.ncbi.nlm.nih.gov/12470132; Raz, A., et al. "Posthypnotic Suggestion and the Modulation of Stroop Interference Under Cycloplegia." *Consciousness and Cognition* 12, no. 3 (September 2003): 332–346. https://pubmed.ncbi.nlm.nih.gov/12941281; MacLeod, C., and P. Sheehan. "Hypnotic Control of Attention in the Stroop Task: A Historical Footnote." *Consciousness and Cognition* 12, no. 3 (September 2003): 347–353. https://pubmed.ncbi.nlm.nih.gov/12941282; Lifshitz, M., et al. "Using Suggestion to Modulate Automatic Processes: From Stroop to McGurk and Beyond." *Cortex* 49, no. 2 (February 2013): 463–473. https://pubmed.ncbi.nlm.nih.gov/23040173; Raz, A., I. Kirsch, and Y. Nitkin-Kaner. "Suggestion Reduces the Stroop Effect." *Psychological Science* 17, no. 2 (February 2006). https://journals.sagepub.com/doi/abs/10.1111/j.1467-9280.2006.01669.x; Raz, A., J. Fan, and M. I. Posner. "Hypnotic Suggestion Reduces Conflict in the Human Brain." *PNAS* 102, no. 28 (June 2005): 9978–9983. https://www.pnas.org/doi/10.1073/pnas.0503064102; Raz, A., M. Moreno-Íñiguez, L. Martin, and H. Zhu. "Suggestion Overrides the Stroop Effect in Highly Hypnotizable Individuals." *Consciousness and Cognition* 16, no. 2 (June 2007): 331–338. https://www.sciencedirect.com/science/article/abs/pii/S1053810006000341

26. Olson, J. A., and A. Raz. "Applying Insights from Magic to Improve Deception in Research: The Swiss Cheese Model." *Journal of Experimental Social Psychology* 92, no. 2 (2020): 104053; Kuhn, G., J. A. Olson, and A. Raz. *The Psychology of Magic and the Magic of Psychology.* Frontiers Media SA, 2016.

27. Hilbig, B. E., I. Thielmann, and R. Böhm. "Bending Our Ethics Code: Avoidable Deception and Its Justification in Psychological Research." *European Psychologist* 27, no. 1 (2021); Reisig, M. D., M. Flippin, and K. Holtfreter. "Toward the Development of a Perceived IRB Violation Scale." *Account Res* 29, no. 5 (2021): 1–15. https://doi.org:10.1080/08989621.2021.1920408; Baumrind, D. "IRBs and Social Science Research: The Costs of Deception." *IRB: Ethics & Human Research* 1, no. 6 (October 1979): 1–4; Maxfield, C. M., et al. "Can the Use of Deception Be Justified in Medical Education Research? A Point/Counterpoint and Case Study." *Academic Radiology* 29, no. 7 (2021). https://doi.org:10.1016/j.acra.2021.05.008

28. Ali, S. S., M. Lifshitz, and A. Raz. "Empirical Neuroenchantment: From Reading Minds to Thinking Critically." *Frontiers in Human Neuroscience* 8, no. 357 (2014): 357. https://doi.org:10.3389/fnhum.2014.00357

29. Ali, S. S., M. Lifshitz, and A. Raz. "Empirical Neuroenchantment: From Reading Minds to Thinking Critically." *Frontiers in Human Neuroscience* 8, no. 357 (2014): 357. https://doi.org:10.3389/fnhum.2014.00357; Theriault, R., S. Veissière, J. A. Olson, and A. Raz. "Treating ADHD with Suggestion-Neurofeedback and Placebo Therapeutics." *Journal of Attention Disorders* 22, no. 8 (June 2018): 707–711; Theriault, R., and A. Raz. "The Psychology of Neurofeedback: Clinical Intervention Even If Applied Placebo." *American Psychologist* 72, no. 7 (2017); Raz, A., and R. Theriault. *Casting Light on the Dark Side of Brain Imaging.* Elsevier Academic Press, 2019; Theriault, R., A. MacPherson, M. Lifshitz, R. R. Roth, and A. Raz. "Neurofeedback with fMRI: A Critical Systematic Review." *Neuroimage* 172 (2018): 786–807.

30. Olson, J. A., M. Landry, K. Appourchaux, and A. Raz. "Simulated Thought Insertion: Influencing the Sense of Agency Using Deception and Magic." *Consciousness and Cognition* 43 (July 2016): 11–26. https://doi.org:https://doi.org/10.1016/j.concog.2016.04.010

31. Olson, J. A., M. Landry, K. Appourchaux, and A. Raz. "Simulated Thought Insertion: Influencing the Sense of Agency Using Deception and Magic." *Consciousness and Cognition* 43 (July 2016): 11–26. https://doi.org:https://doi.org/10.1016/j.concog.2016.04.010

32. Demacheva, I., M. Ladouceur, E. Steinberg, G. Pogossova, and A. Raz. "The Applied Cognitive Psychology of Attention: A Step Closer to Understanding Magic Tricks." *Applied Cognitive Psychology* 26, no. 4 (2012): 541–549. https://doi.org:10.1002/acp.2825

33. Olson, J. A., I. Demacheva, and A. Raz. "Explanations of a Magic Trick Across the Life Span." *Frontiers in Psychology* 6, (March 2015): 219. https://doi.org:10.3389/fpsyg.2015.00219

34. Kelley, J. M., T. J. Kaptchuk, C. Cusin, S. Lipkin, and M. Fava. "Open-Label Placebo for Major Depressive Disorder: A Pilot Randomized Controlled Trial." *Psychotherapy and Psychosomatics* 81, no. 5 (August 2012): 312–314. https://doi.org:10.1159/000337053; Carvalho, C., et al. "Open-Label Placebo Treatment in Chronic Low Back Pain: A Randomized Controlled Trial." *Pain* 157, no. 12 (December 2016): 2766–2772. https://doi.org:10.1097/j.pain.0000000000000700; Schaefer, M., T. Sahin, and B. Berstecher. "Why Do Open-Label Placebos Work? A Randomized Controlled Trial of an Open-Label Placebo Induction with and Without Extended Information About the Placebo Effect in Allergic Rhinitis." *PLOS One* 13, no. 3 (March 2018); Lembo, A., et al. "Open-Label Placebo vs Double-Blind Placebo for Irritable Bowel Syndrome: A Randomized Clinical Trial." *Pain* 162, no. 9 (September 2021): 2428–2435. https://doi.org:10.1097/j.pain.0000000000002234

35. Carvalho, C., et al. "Open-Label Placebo for Chronic Low Back Pain: A 5-Year Follow-Up." *Pain* 162, no. 5 (May 2021): 1521–1527. https://doi.org:10.1097/j.pain.0000000000002162

36. Fassi, L., S. Hochman, Z. J. Daskalakis, D. M. Blumberger, and R. C. Kadosh. "The Importance of Individual Beliefs in Assessing Treatment Efficacy: Insights from Neurostimulation Studies" eLife 12:RP88889, (2023). https://doi.org/10.7554/eLife.88889.2

37. Sandler, A. D., C. E. Glesne, and J. W. Bodfish. "Conditioned Placebo Dose Reduction: A New Teatment in Attention-Deficit Hyperactivity Disorder?" *Journal of Developmental & Behavioral Pediatrics* 31, no. 5 (June 2010): 369–375. https://doi.org:10.1097/DBP.0b013e3181e121ed

38. Benedetti, F. *Placebo Effects: Understanding the Other Side of Medical Care.* 3rd ed. Oxford: Oxford University Press, 2021; Benedetti, F., and D. D. Mitsikostas, *Headache,* edited by P. Martelletti and L. Edvinsson. Springer International Publishing, 2019; Benedetti, F., P. Enck, E. Frisaldi, and M. Schedlowski in *Handbook of Experimental Pharmacology,* edited by M. C. Michel. Springer Berlin, Heidelberg, 2014; Benedetti, F. *Placebo Effects.* 2nd ed. Oxford: Oxford University Press, 2014; Benedetti, F. *Placebo Effects: Understanding the Mechanisms in Health and Disease.* Oxford: Oxford University Press, 2009.

39. Langer, E. *The Mindful Body: Thinking Our Way to Chronic Health.* Ballantine Books, 2023.

40. Fassi, L., S. Hochman, Z. J. Daskalakis, D. M. Blumberger, and R. C. Kadosh. "The Importance of Individual Beliefs in Assessing Treatment Efficacy: Insights from Neurostimulation Studies" eLife 12:RP88889, (2023). https://doi.org/10.7554/eLife.88889.2

Chapter 4

1. Webster, S. T. "Mind over Milkshakes: How Your Mindset Guides Physical Response to Food." *IDEA Health & Fitness* (June 2014): 114.

2. Roberts, A. H., D. G. Kewman, L. Mercier, and M. Hovell. "The Power of Nonspecific Effects in Healing: Implications for Psychosocial and Biological Treatments." *Clinical Psychology Review* 13, no. 5 (1993): 375–391. https://doi.org:10.1016/0272-7358(93)90010-j; Hahn, R. A., and A. Kleinman. "Perspectives of the Placebo Phenomenon: Belief as Pathogen, Belief as Medicine: 'Voodoo Death' and the 'Placebo Phenomenon' in Anthropological Perspective." *Medical Anthropology Quarterly* 14, no. 4 (August 1983): 3–19. https://doi.org:10.1525/maq.1983.14.4.02a00030; Langer, E. *Mindfulness.* Perseus Books Group, 1998.

REFERENCES

3. Rutledge, T. "Calories In, Calories Out: When Weight Loss Psychology Meets Physics." *Psychology Today*, October 16, 2019. https://www.psychologytoday.com/us/blog/the-healthy-journey/201910/calories-in-calories-out

4. Benton, D., and H. A. Young. "Reducing Calorie Intake May Not Help You Lose Body Weight." *Perspectives on Psychological Science* 12, no. 5 (September 2017): 703–714. https://doi.org/10.1177/1745691617690878

5. Finniss, D. G., T. J. Kaptchuk, F. G. Miller, and F. Benedetti. "Biological, Clinical, and Ethical Advances of Placebo Effects." *The Lancet* 375, no. 9715 (February 2010): 686–695. https://doi.org:https://doi.org/10.1016/S0140-6736(09)61706-2; Colloca, L., and F. G. Miller. "Role of Expectations in Health." *Current Opinion in Psychiatry* 24, no. 2 (March 2011): 149–155. https://doi.org:10.1097/YCO.0b013e328343803b; Raz, A., and C. Harris. *Placebo Talks: Modern Perspectives on Placebos in Society.* Oxford: Oxford University Press, 2016.

6. De La Fuente-Fernandez, R., et al. "Expectation and Dopamine Release: Mechanism of the Placebo Effect in Parkinson's Disease." *Science* 293, no. 5532 (August 2001): 1164–1166.

7. Langer, E., M. Djikic, M. Pirson, A. Madenci, and R. Donohue. "Believing Is Seeing: Using Mindlessness (Mindfully) to Improve Visual Acuity." *Psychological Science* 21, no. 5 (May 2010): 661.

8. Raz, A., and C. Harris. *Placebo Talks: Modern Perspectives on Placebos in Society.* Oxford: Oxford University Press, 2016.

9. Shaw, J. E., R. A. Sicree, and P. Z. Zimmet. "Global Estimates of the Prevalence of Diabetes for 2010 and 2030." *Diabetes Research and Clinical Practice* 87, no. 1 (January 2010): 4–14. https://doi.org:10.1016/j.diabres.2009.10.007

10. "Diagnosis and Classification of Diabetes Mellitus." *Diabetes Care* 37, Suppl. 1 (2014): S81–S90. https://doi.org/10.2337/dc14-S081

11. O'Rahilly, S., I. Barroso, and N. J. Wareham. "Genetic Factors in Type 2 Diabetes: The End of the Beginning?" *Science* 307, no. 5708 (January 2005): 370–373. https://doi.org:10.1126/science.1104346

12. Mokdad, A. H., et al. "Prevalence of Obesity, Diabetes, and Obesity-Related Health Risk Factors, 2001." *JAMA* 289, no. 1 (January 2003): 76–79.

13. Stumvoll, M., B. J. Goldstein, and T. W. van Haeften. "Type 2 Diabetes: Principles of Pathogenesis and Therapy." *The Lancet* 365, no. 9467 (April 2005): 1333–1346. https://doi.org/10.1016/S0140-6736(05)61032-X

14. Melanson, K. J., M. S. Westerterp-Plantenga, W. H. Saris, F. J. Smith, and L. A. Campfield. "Blood Glucose Patterns and Appetite in Time-Blinded Humans: Carbohydrate Versus Fat." *American Journal of Physiology* 277, no. 2 (August 1999): R337–R345. https://doi.org/10.1152/ajpregu.1999.277.2.R337

15. Park, C., F. Pagnini, A. Reece, D. Phillips, and E. Langer. "Blood Sugar Level Follows Perceived Time Rather Than Actual Time in People with Type 2 Diabetes." *Proceedings of the National Academy of Sciences of the United States of America* 113, no. 29 (July 2016): 8168–8170. https://doi.org/10.1073/pnas.1603444113

16. Park, C., F. Pagnini, and E. Langer. "Glucose Metabolism Responds to Perceived Sugar Intake More Than Actual Sugar Intake." *Scientific Reports* 10, no. 1 (September 2020). https://doi.org/10.1038/s41598-020-72501-w

17. Park, C., F. Pagnini, A. Reece, D. Phillips, and E. Langer. "Blood Sugar Level Follows Perceived Time Rather Than Actual Time in People with Type 2 Diabetes." *Proceedings of the National Academy of Sciences of the United States of America* 113, no. 29 (July 2016): 8168–8170. https://doi.org/10.1073/pnas.1603444113

18. Sun, F., et al. "Effectiveness of Low-Frequency Pallidal Deep Brain Stimulation at 65 Hz in Tourette Syndrome." *Neuromodulation: Technology at the Neural Interface* 25, no. 2 (February 2022): 286–295. https://doi.org/10.1111/ner.13456; Raz, A. "Recent Perspectives on Tourette's Syndrome: Breakthrough Genetics, a Broadway Musical, and an Educational Documentary." *PsycCRITIQUES* 50, no. 48 (2005). https://doi.org/10.1037/05246512; Morera Maiquez, B. *Investigating Brain Activity Prior to Tics and Modulating This Activity with Peripheral Nerve Stimulation to Suppress Tics in Tourette Syndrome.* PhD thesis, University of Nottingham, 2021.

19. Raz, A., et al. "Neural Substrates of Self-Regulatory Control in Children and Adults with Tourette Syndrome." *Canadian Journal of Psychiatry* 54, no. 9 (September 2009): 579–588. https://doi.org/10.1177/070674370905400902

20. Raz, A. "Translational Attention: From Experiments in the Lab to Helping the Symptoms of Individuals with Tourette's Syndrome." *Consciousness and Cognition* 21, no. 3 (September 2012): 1591–1594. https://doi.org/10.1016/j.concog.2012.05.010

21. "WMA Declaration of Helsinki—Ethical Principles for Medical Research Involving Human Subjects." World Medical Association, October 2022. https://www.wma.net/policies-post/wma-declaration-of-helsinki-ethical-principles-for-medical-research-involving-human-subjects

22. Oluwabusi, O. O., S. Parke, and P. J. Ambrosini. "Tourette Syndrome Associated with Attention Deficit Hyperactivity Disorder: The Impact of Tics and Psychopharmacological Treatment Options." *World Journal of Clinical Pediatrics* 5, no. 1 (February 2016): 128–135. https://doi.org/10.5409/wjcp.v5.i1.128

23. Raz, A., and C. S. Harris. *Placebo Talks: Modern Perspectives on Placebos in Society.* Oxford: Oxford University Press, 2016.

24. Palmieri, J. J., and T. A. Stern. "Lies in the Doctor-Patient Relationship." *The Primary Care Companion to the Journal of Clinical Psychiatry* 11, no. 4 (2009): 163–168. https://doi.org:10.4088/PCC.09r00780

25. Shotton, R. *The Choice Factory: How 25 Behavioural Biases Influence the Products We Decide to Buy.* Harriman House, 2018.

26. Hardwick, M. J., and V. Gruen. *Mall Maker: Victor Gruen, Architect of an American Dream.* University of Pennsylvania Press, 2004.

27. Byrne, R. *The Secret.* New York: Simon & Schuster Audio, 2006.

28. Gawain, S. *Creative Visualization: Use the Power of Your Imagination to Create What You Want in Your Life.* Revised ed. New World Library: Dist. by Publishers Group West, 1995; Gawain, S. *Creative Visualization.* 1st ed. Whatever Pub., 1978.

29. Byrne, R. *The Secret.* New York: Simon & Schuster Audio, 2006.

30. Byrne, R. *The Magic.* Atria Books, 2012.

31. Gibbs, S., I. Harvey, J. C. Sterling, and R. Stark. "Local Treatments for Cutaneous Warts." *Cochrane Database of Systematic Reviews,* July 21, 2003. https://doi.org:10.1002/14651858.CD001781

32. DuBreuil, S. C., and N. P. Spanos in *Handbook of Clinical Hypnosis,* edited by J. W. Rhue, S. J. Lynn, and I. Kirsch, 623–648. American Psychological Association, 1993.

33. DuBreuil, S. C., and N. P. Spanos in *Handbook of Clinical Hypnosis,* edited by J. W. Rhue, S. J. Lynn, and I. Kirsch, 623–648. American Psychological Association, 1993; Sinclair-Gieben, A. H., and D. Chalmers. "Evaluation of Treatment of Warts by Hypnosis." *Lancet* 2, no. 7101 (October 1959): 480–482. https://doi.org:10.1016/s0140-6736(59)90605-1; Tenzel, J. H., and R. L. Taylor. "An Evaluation of Hypnosis and Suggestion as Treatment for Warts." *Psychosomatics* 10, no. 4 (July–August 1969): 252–257. https://doi.org:10.1016/S0033-3182(69)71737-6; French, A. P. "Treatment of Warts by Hypnosis." *American Journal of Obstetrics & Gynecology*

116, no. 6 (July 1973): 887–888. https://doi.org:10.1016/0002-9378(73)91033-8; Surman, O. S., S. K. Gottlieb, T. P. Hackett, and E. L. Silverberg. "Hypnosis in the Treatment of Warts." *Archives of General Psychiatry* 28, no. 3 (March 1973): 439–441. https://doi.org:10.1001/arch psyc.1973.01750330111018; Clawson, T. A., Jr., and R. H. Swade. "The Hypnotic Control of Blood Flow and Pain: The Cure of Warts and the Potential for the Use of Hypnosis in the Treatment of Cancer." *American Journal of Clinical Hypnosis* 17, no. 3 (1975): 160–169. https:// doi.org:10.1080/00029157.1975.10403735; Tasini, M. F., and T. P. Hackett. "Hypnosis in the Treatment of Warts in Immunodeficient Children." *American Journal of Clinical Hypnosis* 19, no. 3 (January 1977): 152–154. https://doi.org:10.1080/00029157.1977.10403862; Dreaper, R. "Recalcitrant Warts on the Hand Cured by Hypnosis." *Practitioner* 220, no. 1316 (1978): 305–310; Johnson, R. F., and T. X. Barber. "Hypnosis, Suggestions, and Warts: An Experimental Investigation Implicating the Importance of 'Believed-in-Efficacy.'" *American Journal of Clinical Hypnosis* 20, no. 3 (January 1978): 165–174. https://doi.org:10.1080/00029157.1978.104039 25; Spanos, N. P., R. J. Stenstrom, and J. C. Johnston. "Hypnosis, Placebo, and Suggestion in the Treatment of Warts." *Psychosomatic Medicine* 50, no. 3 (May 1988): 245–260. https://doi .org:10.1097/00006842-198805000-00003; Ewin, D. M. "Hypnotherapy for Warts (Verruca Vulgaris): 41 Consecutive Cases with 33 Cures." *American Journal of Clinical Hypnosis* 35, no. 1 (July 1992): 1–10. https://doi.org:10.1080/00029157.1992.10402977; Phoenix, S. L. "Psycho-therapeutic Intervention for Numerous and Large Viral Warts with Adjunctive Hypnosis: A Case Study." *American Journal of Clinical Hypnosis* 49, no. 3 (January 2007): 211–218. https://doi .org:10.1080/00029157.2007.10401583

34. Bloch, B. "Über die Heilung der Warzen Durch Suggestion." *Klinische Wochenschrift* 6 (November 1927): 2320–2325. https://doi.org:10.1007/BF01726552

35. Bloch, B. "Über die Heilung der Warzen Durch Suggestion." *Klinische Wochenschrift* 6 (November 1927): 2320–2325. https://doi.org:10.1007/BF01726552

36. Blease, C., L. Colloca, and T. J. Kaptchuk. "Are Open-Label Placebos Ethical? Informed Consent and Ethical Equivocations." *Bioethics* 30, no. 6 (July 2016): 407–414. https://doi.org:10.1111 /bioe.12245

37. Buergler, S., et al. "Imaginary Pills and Open-Label Placebos Can Reduce Test Anxiety by Means of Placebo Mechanisms." *Scientific Reports* 13, no. 1 (February 2023): 2624. https://doi .org:10.1038/s41598-023-29624-7; de Shazer, S. "The Imaginary Pill Technique." *Journal of Strategic and Systemic Therapies* 3, no. 1 (1984): 30–34; Bagge, N., in *Second Official Conference of the Society for Interdisciplinary Placebo Studies (SIPS)*. Leiden, Netherlands, 2019; Wai-Lan Yeung, V., A. L. Geers, and L. Colloca. "Merely Possessing a Placebo Analgesic Improves Analgesia Similar to Using the Placebo Analgesic." *Annals of Behavioral Medicine* 54, no. 9 (September 2020): 637–652. https://doi.org:10.1093/abm/kaaa007

38. Spanos, N. P., R. J. Stenstrom, and J. C. Johnston. "Hypnosis, Placebo, and Suggestion in the Treatment of Warts." *Psychosomatic Medicine* 50, no. 3 (May 1988): 245–260. https://doi .org:10.1097/00006842-198805000-00003; Spanos, N. P., V. Williams, and M. I. Gwynn. "Effects of Hypnotic, Placebo, and Salicylic Acid Treatments on Wart Regression." *Psychosomatic Medicine* 52, no. 1 (January–February 1990): 109–114. https://doi.org:10.1097/00006842 -199001000-00009

39. Castellani, G., T. Croese, J. M. Peralta Ramos, and M. Schwartz. "Transforming the Understand-ing of Brain Immunity." *Science* 380, no. 6640 (April 2023): eabo7649. https://doi.org:10.1126 /science.abo7649

40. Haykin, H., and A. Rolls. "The Neuroimmune Response During Stress: A Physiological Per-spective." *Immunity* 54, no. 9 (September 2021): 1933–1947. https://doi.org:10.1016/j.immu ni.2021.08.023

41. Howe, L. C., J. P. Goyer, and A. J. Crum. "Harnessing the Placebo Effect: Exploring the Influence of Physician Characteristics on Placebo Response: Correction." *Health Psychology* 41, no. 11 (No-vember 2022): 873. https://doi.org:10.1037/hea0001235

REFERENCES

42. Spiegel, D., J. R. Bloom, H. C. Kraemer, and E. Gottheil. "Effect of Psychosocial Treatment on Survival of Patients with Metastatic Breast Cancer." *Lancet* 2, no. 8668 (October 1989): 888–891. https://doi.org:10.1016/s0140-6736(89)91551-1

43. Schwartz, M., and K. Baruch. "Vaccine for the Mind." *Human Vaccines & Immunotherapeutics* 8, no. 10 (October 2012): 1465–1468. https://doi.org:10.4161/hv.21649

44. Baruch, K., and M. Schwartz. "Circulating Monocytes in Between the Gut and the Mind." *Cell Stem Cell* 18, no. 6 (June 2016): 689–691. https://doi.org:10.1016/j.stem.2016.05.008

45. Koren, T., and A. Rolls. "Immunoception: Defining Brain-Regulated Immunity." *Neuron* 110, no. 21 (November 2022): 3425–3428. https://doi.org:10.1016/j.neuron.2022.10.016

46. Everson-Rose, S. A., and T. T. Lewis. "Psychosocial Factors and Cardiovascular Diseases." *Annual Review of Public Health* 26, (2005): 469–500. https://doi.org:10.1146/annurev .publhealth.26.021304.144542

47. Langer, E. *The Mindful Body: Thinking Our Way to Chronic Health.* Ballantine Books, 2023.

48. Ben-Shaanan, T. L., et al. "Modulation of Anti-Tumor Immunity by the Brain's Reward System." *Nature Communications* 9, no. 2723 (2018): 2723. https://doi.org:10.1038/s41467-018-05283-5

49. Li, S. B., et al. "Hypothalamic Circuitry Underlying Stress-Induced Insomnia and Peripheral Immunosuppression." *Science Advances* 6, no. 37 (September 2020): eabc2590. https://doi .org:10.1126/sciadv.abc2590

50. Koren, T., et al. "Insular Cortex Neurons Encode and Retrieve Specific Immune Responses." *Cell* 184, no. 24 (November 2021): 6211. https://doi.org:10.1016/j.cell.2021.11.021

51. Palsson, O. S. "Standardized Hypnosis Treatment for Irritable Bowel Syndrome: The North Carolina Protocol." *International Journal of Clinical and Experimental Hypnosis* 54, no. 1 (January 2006): 51–64. https://doi.org:10."1080/00207140500322933

52. Lembo, A., et al. "Open-Label Placebo vs Double-Blind Placebo for Irritable Bowel Syndrome: A Randomized Clinical Trial." *Pain* 162, no. 9 (September 2021): 2428–2435. https://doi .org:10.1097/j.pain.0000000000002234; Nurko, S., et al. "Effect of Open-Label Placebo on Children and Adolescents with Functional Abdominal Pain or Irritable Bowel Syndrome: A Randomized Clinical Trial." *JAMA Pediatrics* 176, no. 4 (April 2022): 349–356. https://doi.org:10.1001 /jamapediatrics.2021.5750; Ballou, S., et al. "Psychological Predictors of Response to Open-Label Versus Double-Blind Placebo in a Randomized Controlled Trial in Irritable Bowel Syndrome." *Psychosomatic Medicine* 84, no. 6 (July–August 2022): 738–746. https://doi.org:10.1097 /PSY.0000000000001078; Ballou, S., et al. "Open-Label Versus Double-Blind Placebo Treatment in Irritable Bowel Syndrome: Study Protocol for a Randomized Controlled Trial." *Trials* 18, no. 1 (May 2017): 234. https://doi.org:10.1186/s13063-017-1964-x

53. Poller, W. C., et al. "Brain Motor and Fear Circuits Regulate Leukocytes During Acute Stress." *Nature* 607, no. 7919 (July 2022): 578–584. https://doi.org:10.1038/s41586-022-04890-z

54. Marilov, V. V. "[A Transition of a Functional Disease to an Organic Psychosomatic Disorder]." *Zh Nevrol Psikhiatr Im S S Korsakova* 106, no. 1 (2006): 21–23.

55. Paolicelli, R. C., et al. "Microglia States and Nomenclature: A Field at Its Crossroads." *Neuron* 110, no. 21 (November 2022): 3458–3483. https://doi.org:10.1016/j.neuron.2022.10.020

56. Castellani, G., T. Croese, J. M. Peralta Ramos, and M. Schwartz. "Transforming the Understanding of Brain Immunity." *Science* 380, no. 6640 (April 2023): eabo7649. https://doi.org:10.1126 /science.abo7649

57. Schwartz, M., M. A. Abellanas, A. Tsitsou-Kampeli, and S. Suzzi. "The Brain-Immune Ecosystem: Implications for Immunotherapy in Defeating Neurodegenerative Diseases." *Neuron* 110, no. 21 (November 2022): 3421–3424. https://doi.org:10.1016/j.neuron.2022.09.007

Chapter 5

1. Shorvon, S., E. Perucca, and J. Engel, eds. *The Treatment of Epilepsy.* 4th ed. Wiley Blackwell, 2016.

2. Kavic, M. S. "The US Medical Liability System—A System in Crisis." *Journal of the Society of Laparoscopic & Robotic Surgeons* 8, no. 1 (January–March 2004): 1–2; Hauser, M. J., M. L. Commons, H. J. Bursztajn, and T. G. Gutheil. "Fear of Malpractice Liability and Its Role in Clinical Decision Making." In *Decision Making in Psychiatry and the Law,* edited by T. G. Gutheil, H. J. Bursztajn, A. Brodsky, and V. Alexander, 209–226. Williams & Wilkins Co., 1991; Tolchin, B. Personal Communication. 2020.

3. Shostak, S., and R. Ottman. "Ethical, Legal, and Social Dimensions of Epilepsy Genetics." *Epilepsia* 47, no. 10 (October 2006): 1595–1602; Soyer, O., U. M. Sahiner, and B. E. Sekerel. "Pro and Contra: Provocation Tests in Drug Hypersensitivity." *International Journal of Molecular Sciences* 18, no. 7 (July 2017): 1437. https://doi.org:10.3390/ijms18071437

4. Britton, J. W., et al. *Electroencephalography (EEG): An Introductory Text and Atlas of Normal and Abnormal Findings in Adults, Children, and Infants.* Chicago: American Epilepsy Society, 2016; Bomela, W., S. Wang, C. A. Chou, and J. S. Li. "Real-Time Inference and Detection of Disruptive EEG Networks for Epileptic Seizures." *Scientific Reports* 10, no. 1 (May 2020): 8653. https://doi.org:10.1038/s41598-020-65401-6

5. McLane, H. C., et al. "Availability, Accessibility, and Affordability of Neurodiagnostic Tests in 37 Countries." *Neurology* 85, no. 18 (November 2015): 1614–1622.

6. Surges, R. Personal Communication. University of Bonn; Reuber, M. Personal Communication. *Seizure European Journal of Epilepsy* (2020).

7. Clinical Epilepsy.

8. Roepstorff, A., and C. Frith. "What's at the Top in the Top-Down Control of Action? Script-Sharing and 'Top-Top' Control of Action in Cognitive Experiments." *Psychological Research* 68, no. 2–3 (April 2004): 189–198. https://doi.org:10.1007/s00426-003-0155-4

9. McKinley, G. P., R. Darnell, D. Jacobs, and N. Williams. "Nibi and Cultural Affordance at Walpole Island First Nations: Environmental Change and Mental Health." *International Journal of Environmental Research and Public Health* 19, no. 14 (July 2022): 8623. https://doi.org:10.3390/ijerph19148623; Kavousipor, S., et al. "Cultural Adaptation and Psychometric Properties of the Persian Version of the Affordance in the Home Environment for Motor Development." *Iranian Journal of Child Neurology* 13, no. 1 (2019): 25–35; Larkin, M., et al. "Cultural Affordance, Social Relationships, and Narratives of Independence: Understanding the Meaning of Social Care for Adults with Intellectual Disabilities from Minority Ethnic Groups in the UK." *International Journal of Developmental Disabilities* 64, no. 3 (2018): 195–203. https://doi.org:10.1080/2047386 9.2018.1469807; Ramstead, M. J. D., S. P. L. Veissière, and L. J. Kirmayer. "Cultural Affordances: Scaffolding Local Worlds Through Shared Intentionality and Regimes of Attention." *Frontiers in Psychology* 7, (2016): 1090. https://doi.org:10.3389/fpsyg.2016.01090

10. Ramstead, M. J. D., S. P. L. Veissière, and L. J. Kirmayer. "Cultural Affordances: Scaffolding Local Worlds Through Shared Intentionality and Regimes of Attention." *Frontiers in Psychology* 7, (2016): 1090. https://doi.org:10.3389/fpsyg.2016.01090

11. Benbadis, S. Personal Communication. 2020.

12. Mackenzie, J. N. "The Production of the So-Called 'Rose Cold' by Means of an Artificial Rose." *The American Journal of the Medical Sciences* 91, no. 181 (January 1886): 45.

13. Ross, A., and D. Fleming. "Hayfever—Practical Management Issues." *British Journal of General Practice* 54, no. 503 (June 2004): 412–414.

14. Bateman, E. D., et al. "Global Strategy for Asthma Management and Prevention: GINA Executive Summary." *European Respiratory Journal* 31, no. 1 (January 2008): 143–178. https://doi.org:10.1183/13993003.51387-2007; Lougheed, M. D., et al. "Canadian Thoracic Society

REFERENCES

2012 Guideline Update: Diagnosis and Management of Asthma in Preschoolers, Children, and Adults." *Canadian Respiratory Journal* 19, no. 2 (March–April 2012): 127–164. https:// doi.org:10.1155/2012/635624; Lemiere, C., et al. "Adult Asthma Consensus Guidelines Update 2003." *Canadian Respiratory Journal* 11, suppl. A (May 2004): 9A–18A. https://doi .org:10.1155/2004/271362; "Expert Panel Report 3 (EPR-3): Guidelines for the Diagnosis and Management of Asthma—Summary Report 2007." *Journal of Allergy and Clinical Immunology* 120, suppl. 5 (November 2007): S94–138. https://doi.org:10.1016/j.jaci.2007.09.043; Tarlo, S. M., et al. "Canadian Thoracic Society Guidelines for Occupational Asthma." *Canadian Respiratory Journal* 5, no. 4 (July–August 1998): 289–300. https://doi.org:10.1155/1998/587580

15. Delpero, E. A., E. Turkakin, and A. Raz. "Asthma as a Model for Placebo Effects in Modern Medicine." *The Journal of Mind-Body Regulation* 1, no. 3 (October 2011): 161–168; Hackman, R. M., J. S. Stern, and M. E. Gershwin. "Hypnosis and Asthma: A Critical Review." *Journal of Asthma* 37, no. 1 (February 2000): 1–15. https://doi.org:10.3109/02770900009055424

16. Kaptchuk, T. J., et al. "Placebos Without Deception: A Randomized Controlled Trial in Irritable Bowel Syndrome." *PLOS One* 5, no. 12 (December 2010): e15591. https://doi.org:10.1371 /journal.pone.0015591; Wechsler, M. E., et al. "Active Albuterol or Placebo, Sham Acupuncture, or No Intervention in Asthma." *New England Journal of Medicine* 365, no. 2 (July 2011): 119– 126; Kaptchuk, T. J., et al. "Do 'Placebo Responders' Exist?" *Contemporary Clinical Trials* 29, no. 4 (July 2008): 587–595. https://doi.org:10.1016/j.cct.2008.02.002; Sodergren, S. C., and M. E. Hyland. "Expectancy and Asthma." *American Psychological Association* (1999): 197–212. https:// doi.org: https://doi.org/10.1037/10332-008

17. Romero-Velarde, E., et al. "The Importance of Lactose in the Human Diet: Outcomes of a Mexican Consensus Meeting." *Nutrients* 11, no. 11 (November 2019): 2737. https://doi.org:10.3390 /nu11112737

18. Lapides, R. A., and D. A. Savaiano. "Gender, Age, Race and Lactose Intolerance: Is There Evidence to Support a Differential Symptom Response? A Scoping Review." *Nutrients* 10, no. 12 (December 2018): 1956. https://doi.org:10.3390/nu10121956

19. Blinderman, I. "Mind Over Matter in Biological Conditions—The Role of Psychological Processes in Lactose Intolerance and Chronic Spontaneous Urticaria." Academic thesis, McGill University, 2012.

20. Kihlstrom, J. F. "Prospects for De-Automatization." *Consciousness and Cognition* 20, no. 2 (2011): 332–334. https://doi.org:10.1016/j.concog.2010.03.004; Kihlstrom, J. F. "Neuro-Hypnotism: Prospects for Hypnosis and Neuroscience." *Cortex* 49, no. 2 (February 2013): 365–374. https:// doi.org:10.1016/j.cortex.2012.05.016

21. Azizi, M., and F. Elyasi. "Biopsychosocial View to Pseudocyesis: A Narrative Review." *International Journal of Reproductive Biomedicine* 15, no. 9 (September 2017): 535–542.

22. Gaskin, I. M. "Has Pseudocyesis Become an Outmoded Diagnosis?" *Birth: Issues in Perinatal Care* 39, no. 1 (March 2012): 77–79. https://doi.org:10.1111/j.1523-536X.2011.00521.x

23. Bianchi-Demicheli, F., F. Lüdicke, and D. Chardonnens. "Imaginary Pregnancy 10 Years After Abortion and Sterilization in a Menopausal Woman: A Case Report." *Maturitas* 48, no. 4 (August 2004): 479–481. https://doi.org:10.1016/j.maturitas.2003.09.030; Yadav, T., Y. P. Balhara, and D. K. Kataria. "Pseudocyesis Versus Delusion of Pregnancy: Differential Diagnoses to Be Kept in Mind." *Indian Journal of Psychological Medicine* 34, no. 1 (January–March 2012): 82–84. https:// doi.org:10.4103/0253-7176.96167

24. Tarín, J. J., C. Hermenegildo, M. A. García-Pérez, and A. Cano. "Endocrinology and Physiology of Pseudocyesis." *Reproductive Biology and Endocrinology* 11, (May 2013): 39. https://doi .org:10.1186/1477-7827-11-39

25. O'Grady, J. P., and M. Rosenthal. "Pseudocyesis: A Modern Perspective on an Old Disorder." *Obstetrical & Gynecological Survey* 44, no. 7 (July 1989): 500–511; Simon, M., V. Voros, R. Herold, S. Fekete, and T. Tenyi. "Delusions of Pregnancy with Post-Partum Onset: An Integrated, Individualized View." *European Journal of Psychiatry* 23, no. 4 (October–December 2009).

REFERENCES

26. Del Pizzo, J., L. Posey-Bahar, and R. Jimenez. "Pseudocyesis in a Teenager with Bipolar Disorder." *Clinical Pediatrics* 50, no. 2 (February 2011): 169–171. https://doi .org:10.1177/0009922809350675; Craddock, B., N. Craddock, and L. I. Liebling. "Pseudocyesis Followed by Depressive Psychosis." *British Journal of Psychiatry* 157, (October 1990): 624–625. https://doi.org:10.1192/bjp.157.4.624a

27. Miller, W. H., and R. Maricle. "Pseudocyesis: A Model for Cultural, Psychological, and Biological Interplay." *Journal of Psychosomatic Obstetrics & Gynecology* 8, no. 3 (1988): 183–190. https://doi .org:10.3109/01674828809016786

28. Yang, J. J., C. M. Larsen, D. R. Grattan, and M. S. Erskine. "Mating-Induced Neuroendocrine Responses During Pseudopregnancy in the Female Mouse." *Journal of Neuroendocrinology* 21, no. 1 (January 2009): 30–39. https://doi.org:10.1111/j.1365-2826.2008.01803.x; van Kesteren, F., et al. "The Physiology of Cooperative Breeding in a Rare Social Canid; Sex, Suppression and Pseudopregnancy in Female Ethiopian Wolves." *Physiology & Behavior* 122 (October 2013): 39–45. https://doi.org:10.1016/j.physbeh.2013.08.016; Northrop, L. E., J. L. Shadrach, and M. S. Erskine. "Noradrenergic Innervation of the Ventromedial Hypothalamus Is Involved in Mating-Induced Pseudopregnancy in the Female Rat." *Journal of Neuroendocrinology* 18, no. 8 (August 2006): 577–583. https://doi.org:10.1111/j.1365-2826.2006.01453.x; Guedes, D., and R. J. Young. "A Case of Pseudo-Pregnancy in Captive Brown Howler Monkeys (Alouatta Guariba)." *Folia Primatologica* 75, no. 5 (September–October 2004): 335–338. https://doi .org:10.1159/000080211

29. Campos, S. J., and D. Link. "Pseudocyesis." *The Journal for Nurse Practitioners* 12, no. 6 (June 2016): 390–394. https://doi.org:10.1016/j.nurpra.2016.03.009

30. Bivin, G. D., and M. P. Klinger. *Pseudocyesis*. Principia Press, 1937.

31. Grover, S., A. Sharma, D. Ghormode, and N. Rajpal. "Pseudocyesis: A Complication of Antipsychotic-Induced Increased Prolactin Levels and Weight Gain." *Journal of Pharmacology & Pharmacotherapeutics* 4, no 3 (July–September 2013); Tarín, J. J., C. Hermenegildo, M. A. García-Pérez, and A. Cano. "Endocrinology and Physiology of Pseudocyesis." *Reproductive Biology and Endocrinology* 11, (May 2013): 39. https://doi.org:10.1186/1477-7827-11-39

32. Kenner, W. D., and S. E. Nicolson. "Psychosomatic Disorders of Gravida Status: False and Denied Pregnancies." *Psychosomatics: Journal of Consultation and Liaison Psychiatry* 56, no. 2 (March–April 2015): 119–128. https://doi.org:10.1016/j.psym.2014.09.004

33. Azizi, M., and F. Elyasi. "Biopsychosocial View to Pseudocyesis—A Narrative Review." *International Journal of Reproductive Biomedicine* 15, no. 9 (2017): 535–542.

34. Steinberg, A., et al. "Psychoendocrine Relationships in Pseudocyesis." *Psychosomatic Medicine* 8, (May–June 1946): 176–179.

35. Marušić, S., D. Karlović, Z. Zoričić, M. Martinac, and L. Jokanović. "Pseudocyesis: A Case Report." *Acta Clinica Croatica* 44, no. 3 (2005).

36. Mortimer, A., and J. Banbery. "Pseudocyesis Preceding Psychosis." *British Journal of Psychiatry* 152, (1988): 560–562. https://doi.org:10.1192/bjp.152.4.562

37. Greaves, D. C., P. E. Green, and L. J. West. "Psychodynamic and Psychophysiological Aspects of Pseudocyesis." *Psychosomatic Medicine* 22, no. 1 (January 1960): 24–31. https://doi.org :10.1097/00006842-196001000-00004; Ibekwe, P. C., and J. U Achor. "Psychosocial and Cultural Aspects of Pseudocyesis." *Indian Journal of Psychiatry* 50, no. 2 (April–June 2008): 112–116.

38. Soffer-Dudek, N. "Dissociative Absorption, Mind-Wandering, and Attention-Deficit Symptoms: Associations with Obsessive-Compulsive Symptoms." *British Journal of Clinical Psychology* 58, no. 1 (2019): 51–69. https://doi.org:10.1111/bjc.12186

39. Bob, P., and O. Louchakova. "Dissociative States in Dreams and Brain Chaos: Implications for Creative Awareness." *Frontiers in Psychology* 6, (2015): 1353. https://doi.org:10.3389 /fpsyg.2015.01353

REFERENCES

40. Pearson, D. G., C. Deeprose, S. M. Wallace-Hadrill, S. Burnett Heyes, and E. A. Holmes. "Assessing Mental Imagery in Clinical Psychology: A Review of Imagery Measures and a Guiding Framework." *Clinical Psychology Review* 33, no. 1 (February 2013): 1–23. https://doi.org:10.1016/j.cpr.2012.09.001

41. Raz, A., and M. Lifshitz. *Hypnosis and Meditation: Towards an Integrative Science of Conscious Planes.* Oxford: Oxford University Press, 2016.

42. Luhrmann, T. M. *When God Talks Back: Understanding the American Evangelical Relationship with God.* Reprint ed. Vintage, 2012; *Hearing the Voice: Interdisciplinary Voice-Hearing Research* (2021). https://hearingthevoice.org/a-z-list-of-publications

43. DSM-5 Mental Disorders; Krystal, J. H., et al. "Subanesthetic Effects of the Noncompetitive NMDA Antagonist, Ketamine, in Humans. Psychotomimetic, Perceptual, Cognitive, and Neuroendocrine Responses." *Archives of General Psychiatry* 51, no. 3 (March 1994): 199–214. https://doi.org:10.1001/archpsyc.1994.03950030035004; Ferezou, I., et al. "Spatiotemporal Dynamics of Cortical Sensorimotor Integration in Behaving Mice." *Neuron* 56, no. 5 (December 2007): 907–923. https://doi.org:10.1016/j.neuron.2007.10.007; Mohajerani, M. H., et al. "Spontaneous Cortical Activity Alternates Between Motifs Defined by Regional Axonal Projections." *Nature Neuroscience* 16, no. 10 (2013): 1426–1435. https://doi.org:10.1038/nn.3499; Guo, Z. V., et al. "Flow of Cortical Activity Underlying a Tactile Decision in Mice." *Neuron* 81, no. 1 (January 2014): 179–194. https://doi.org:10.1016/j.neuron.2013.10.020; Ma, Y., et al. "Resting-State Hemodynamics Are Spatiotemporally Coupled to Synchronized and Symmetric Neural Activity in Excitatory Neurons." *Proceedings of the National Academy of Sciences of the United States of America* 113, no. 52 (2016): E8463–E8471. https://doi.org:10.1073/pnas.1525369113; Wekselblatt, J. B., E. D. Flister, D. M. Piscopo, and C. M. Niell. "Large-Scale Imaging of Cortical Dynamics During Sensory Perception and Behavior." *Journal of Neurophysiology* 115, no. 6 (June 2016): 2852–2866. https://doi.org:10.1152/jn.01056.2015; Allen, W. E., et al. "Global Representations of Goal-Directed Behavior in Distinct Cell Types of Mouse Neocortex." *Neuron* 94, no. 4 (May 2017): 891–907 e896. https://doi.org:10.1016/j.neuron.2017.04.017; Chen, T. W., N. Li, K. Daie, and K. Svoboda. "A Map of Anticipatory Activity in Mouse Motor Cortex." *Neuron* 94, no. 4 (May 2017): 866–879.e4. https://doi.org:10.1016/j.neuron.2017.05.005; Makino, H., et al. "Transformation of Cortex-Wide Emergent Properties During Motor Learning." *Neuron* 94, no. 4 (May 2017): 880–890.e8. https://doi.org:10.1016/j.neuron.2017.04.015; Xiao, D., et al. "Mapping Cortical Mesoscopic Networks of Single Spiking Cortical or Sub-Cortical Neurons." *Elife* 6, (2017). https://doi.org:10.7554/eLife.19976; Gilad, A., Y. Gallero-Salas, D. Groos, and F. Helmchen. "Behavioral Strategy Determines Frontal or Posterior Location of Short-Term Memory in Neocortex." *Neuron* 99, no. 4 (August 2018): 814–828.e7. https://doi.org:10.1016/j.neuron.2018.07.029; Musall, S., M. T. Kaufman, A. L. Juavinett, S. Gluf, and A. K. Churchland. "Single-Trial Neural Dynamics Are Dominated by Richly Varied Movements." *Nature Neuroscience* 22, no. 10 (October 2019): 1677–1686. https://doi.org:10.1038/s41593-019-0502-4; Kauvar, I. V., et al. "Cortical Observation by Synchronous Multifocal Optical Sampling Reveals Widespread Population Encoding of Actions." *Neuron* 107, no. 2 (July 2020): 351–367.e19. https://doi.org:10.1016/j.neuron.2020.04.023

44. Grant, J. A., et al. "Cortical Thickness, Mental Absorption and Meditative Practice: Possible Implications for Disorders of Attention." *Biology Psychology* 92, no. 2 (February 2013): 275–281. https://doi.org:10.1016/j.biopsycho.2012.09.007; Bruckmaier, M., I. Tachtsidis, P. Phan, and N. Lavie. "Attention and Capacity Limits in Perception: A Cellular Metabolism Account." *Journal of Neuroscience* 40, no. 35 (August 2020): 6801–6811. https://doi.org:10.1523/JNEUROSCI.2368-19.2020

45. Seligman, R., and L. J. Kirmayer. "Dissociative Experience and Cultural Neuroscience: Narrative, Metaphor, and Mechanism." *Culture, Medicine, and Psychiatry* 32, no. 1 (March 2008): 31–64. https://doi.org:10.1007/s11013-007-9077-8

46. Carlson, E. B., C. Dalenberg, and E. McDade-Montez. "Dissociation in Posttraumatic Stress Disorder Part I: Definitions and Review of Research." *Psychological Trauma: Theory, Research, Practice, and Policy* 4, no. 5 (2012): 479–489. https://doi.org:10.1037/a0027748; Medford, N. "Dissociative Symptoms and Epilepsy." *Epilepsy & Behavior* 30, (January 2014).

47.   Najavits, L. M., and M. Walsh. "Dissociation, PTSD, and Substance Abuse: An Empirical Study." *Journal of Trauma & Dissociation* 13, no. 1 (2012): 115–126. https://doi.org:10.1080/1529973 2.2011.608781; Schifano, F., F. Napoletano, S. Chiappini, et al. "New Psychoactive Substances (NPS), Psychedelic Experiences and Dissociation: Clinical and Clinical Pharmacological Issues." *Current Addiction Reports* 6, (2019): 140–152.

48.   Vesuna, S., et al. "Deep Posteromedial Cortical Rhythm in Dissociation." *Nature* 586, no. 7827 (October 2020): 87–94. https://doi.org:10.1038/s41586-020-2731-9

Chapter 6

1.   Raaijmakers, J. G. W. "Hypnosis, Will, and Memory: A Psycholegal History." (1990); Loftus, E. F. *Witness for the Defense: The Accused, the Eyewitnesses, and the Expert Who Puts Memory on Trial.* St. Martin's Press, 1991.

2.   Aviv, R. "How Elizabeth Loftus Changed the Meaning of Memory." *New Yorker*, March 2021.

3.   Loftus, E. F. *Witness for the Defense: The Accused, the Eyewitnesses, and the Expert Who Puts Memory on Trial.* St. Martin's Press, 1991.

4.   Loftus, E. "How Reliable Is Your Memory?" TEDGlobal. June 2013. https://www.ted.com/talks /elizabeth_loftus_how_reliable_is_your_memory?utm_campaign=tedspread&utm_medium=refer ral&utm_source=tedcomshare

5.   Loftus, E. F. "Leading Questions and the Eyewitness Report." *Cognitive Psychology* 7, no. 4 (October 1975): 560–572.

6.   Loftus, E. F. "Searching for the Neurobiology of the Misinformation Effect." *Learning & Memory* 12, no. 1 (January 2005): 1–2. https://doi.org:10.1101/lm.90805

7.   Cochran, K. J., R. L. Greenspan, D. F. Bogart, and E. F. Loftus. "Memory Blindness: Altered Memory Reports Lead to Distortion in Eyewitness Memory." *Memory & Cognition* 44, no. 5 (July 2016): 717–726.

8.   Toglia, M. P., D. F. Ross, J. Pozzulo, and E. Pica. *The Elderly Eyewitness in Court.* Psychology Press, 2014.

9.   Loftus, E. F., and H. G. Hoffman. "Misinformation and Memory: The Creation of New Memories." *Journal of Experimental Psychology: General* 118, no. 1 (1989): 100–104.

10.   Loftus, E. F., and J. C. Palmer. "Reconstruction of Automobile Destruction: An Example of the Interaction Between Language and Memory." *Journal of Verbal Learning and Verbal Behavior* 13, no. 5 (1974): 585–589.

11.   Loftus, E. F. "Leading Questions and the Eyewitness Report." *Cognitive Psychology* 7, no. 4 (October 1975): 560–572; Loftus, E. F., and H. G. Hoffman. "Misinformation and Memory: The Creation of New Memories." *Journal of Experimental Psychology: General* 118, no. 1 (1989): 100–104; Loftus, E. F., and J. C. Palmer. "Reconstruction of Automobile Destruction: An Example of the Interaction Between Language and Memory." *Journal of Verbal Learning and Verbal Behavior* 13, no. 5 (1974): 585–589; Zhu, B., et al. "Individual Differences in False Memory from Misinformation: Cognitive Factors." *Memory* 18, no. 5 (July 2010): 543–555; Weingardt, K. R., E. F. Loftus, and D. S. Lindsay. "Misinformation Revisited: New Evidence on the Suggestibility of Memory." *Memory & Cognition* 23, no. 1 (January 1995): 72–82. https://doi.org:10.3758/BF03210558; Pohl, R. *Cognitive Illusions: A Handbook on Fallacies and Biases in Thinking, Judgement and Memory.* Psychology Press, 2004; Castelle, G., and E. F. Loftus. "Misinformation and Wrongful Convictions." In *Wrongly Convicted: Perspectives on Failed Justice,* edited by S. D. Westervelt and J. A. Humphrey, 17–35. New Brunswick: Rutgers University Press, 2001; Braun, K. A., and E. F. Loftus. "Advertising's Misinformation Effect." *Applied Cognitive Psychology* 12, no. 6 (December 1998): 569–591; Belli, R. F., and E. F. Loftus in *Remembering Our Past: Studies in Autobiographical Memory,* edited by D. C. Rubin, 157–179. Cambridge University Press, 1996.

12. Loftus, E. F. "Lost in the Mall: Misrepresentations and Misunderstandings." *Ethics & Behavior* 9, no. 1 (1999): 51–60. https://doi.org:10.1207/s15327019eb0901_4

13. Kloft, L., et al. "Cannabis Increases Susceptibility to False Memory." *Proceedings of the National Academy of Sciences* 117, no. 9 (February 2020): 4584–4589.

14. Lynn, S. J., T. Lock, E. F. Loftus, E. Krackow, and S. O. Lilienfeld. "The Remembrance of Things Past: Problematic Memory Recovery Techniques in Psychotherapy." In *Science and Pseudoscience in Clinical Psychology*, edited by S. O. Lilienfeld, J. M. Lohr, and S. J. Lynn, 205–239. Guilford Press, 2003; Lynn, S. J., E. F. Loftus, S. O. Lilienfeld, and T. Lock. "Recovery Techniques in Psychotherapy: Problems and Pitfalls." *Skeptical Inquirer* (July–August 2003): 40–46; Loftus, E. F., and M. Yapko. "Psychotherapy and the Recovery of Repressed Memories." In *True and False Allegations of Child Sexual Abuse: Assessment and Case Management*, edited by T. Ney, 176–191. Brunner/Mazel, 1995; Loftus, E. F., E. M. Milo, and J. R. Paddock. "The Accidental Executioner: Why Psychotherapy Must Be Informed by Science." *Counseling Psychologist* 12, no. 2 (1995): 300–309. https://doi.org:10.1177/0011000095232005

15. LePort, A. K. R., et al. "Behavioral and Neuroanatomical Investigation of Highly Superior Autobiographical Memory (HSAM)." *Neurobiology of Learning and Memory* 98, no. 1 (July 2012): 78–92. https://doi.org:10.1016/j.nlm.2012.05.002

16. Raz, A., et al. "A Slice of π: An Exploratory Neuroimaging Study of Digit Encoding and Retrieval in a Superior Memorist." *Neurocase* 15, no. 5 (October 2009): 361–372. https://doi.org:10.1080/13554790902776896

17. Parker, E. S., L. Cahill, and J. L. McGaugh. "A Case of Unusual Autobiographical Remembering." *Neurocase* 12, no. 1 (February 2006): 35–49. https://doi.org:10.1080/13554790500473680

18. Lawrence, P., et al. "False Memories in Highly Superior Autobiographical Memory Individuals." *Proceedings of the National Academy of Sciences of the United States of America* 110, no. 52 (November 2013): 20947–20952.

19. Yu, J., Q. Tao, R. Zhang, C. C. H. Chan, and T. M. C. Lee. "Can fMRI Discriminate Between Deception and False Memory? A Meta-Analytic Comparison Between Deception and False Memory Studies." *Neuroscience & Behavioral Reviews* 104, (September 2019): 43–55. https://doi.org/10.1016/j.neubiorev.2019.06.027; Kurkela, K. A., and N. A. Dennis. "Event-Related fMRI Studies of False Memory: An Activation Likelihood Estimation Meta-Analysis." *Neuropsychologia* 81, (January 2016): 149–167. https://doi.org/10.1016/j.neuropsychologia.2015.12.006; Ayat, N., et al. "An Extension of 'Can fMRI Discriminate Between Deception and False Memory? A Meta-Analytic Comparison Between Deception and False Memory Studies.'" *UC Berkeley: Cognitive Science & Psychology Division, ULAB*, 2020.

20. Shaw, J. "Do False Memories Look Real? Evidence That People Struggle to Identify Rich False Memories of Committing Crime and Other Emotional Events." *Frontiers in Psychology* 11, (April 2020): 650. https://doi.org:10.3389/fpsyg.2020.00650; Campbell, M. A., and S. Porter. "Pinpointing Reality: How Well Can People Judge True and Mistaken Emotional Childhood Memories?" *Canadian Journal of Behavioural Science* 34, no. 4 (2002): 217–229.

21. Wing, E. A., et al. "Cortical Overlap and Cortical-Hippocampal Interactions Predict Subsequent True and False Memory." *The Journal of Neuroscience* 40, no. 9 (February 2020): 1920–1930. https://doi.org:10.1523/JNEUROSCI.1766-19.2020

22. Ye, Z., et al. "Neural Global Pattern Similarity Underlies True and False Memories." *The Journal of Neuroscience* 36, no. 25 (June 2016): 6792–6802. https://doi.org:10.1523/JNEUROSCI.0425-16.2016; Bernstein, D. M., and E. F. Loftus. "How to Tell If a Particular Memory Is True or False." *Perspectives on Psychological Science* 4, no. 4 (July 2009): 370–374.

23. Raz, A., and R. T. Thibault. *Casting Light on the Dark Side of Brain Imaging*. Academic Press, an imprint of Elsevier, 2019.

24. Loftus, E. "How Reliable Is Your Memory?" TEDGlobal. June 2013. https://www.ted.com /talks/elizabeth_loftus_how_reliable_is_your_memory?utm_campaign=tedspread&utm_medium =referral&utm_source=tedcomshare

25. Stephen, M. N., and S. Itamar. "Attribute-Task Compatibility as a Determinant of Consumer Preference Reversals." *Journal of Marketing Research* 34, no. 2 (1997): 205–218. https://doi .org:10.2307/3151859; Schwarz, N. E., and S. E. Sudman. *Answering Questions: Methodology for Determining Cognitive and Communicative Processes in Survey Research.* Jossey-Bass, 1996; Feldman, J. M., and J. G. Lynch, Jr. "Self-Generated Validity and Other Effects of Measurement on Belief, Attitude, Intention, and Behavior." *Journal of Applied Psychology* 73, no. 3 (1988): 421–435; Fischer, G. W., and S. A. Hawkins. "Strategy Compatibility, Scale Compatibility, and the Prominence Effect." *Journal of Experimental Psychology: Human Perception and Performance* 19, no. 3 (1993): 580–597; Coupey, E., J. R. Irwin, and J. W. Payne. "Product Category Familiarity and Preference Construction." *Journal of Consumer Research* 24, no. 4 (March 1998): 459–468. https://doi.org:10.1086/209521; Simmons, C. J., B. A. Bickart, and J. G. Lynch, Jr. "Capturing and Creating Public Opinion in Survey Research." *Journal of Consumer Research* 20, no. 2 (September 1993): 316–329.

26. Davis, R. H. "The Anatomy of a Smear Campaign." *Boston Globe*, March 21, 2004.

27. Gross, S. R. "Determining the Neutrality of Death-Qualified Juries: Judicial Appraisal of Empirical Data." *Law and Human Behavior* 8, no. 1–2 (1984): 7–30.

28. Davis, R. H. "The Anatomy of a Smear Campaign." *Boston Globe*, March 21, 2004.

29. Moore, S. G., D. T. Neal, G. J. Fitzsimons, and B. Shiv. "Wolves in Sheep's Clothing: How and When Hypothetical Questions Influence Behavior." *Organizational Behavior and Human Decision Processes* 117, no. 1 (2012): 168–178. https://doi.org/10.1016/j.obhdp.2011.08.003

30. Kuhn, G., and A. Pailhes. "Influencing Choices with Conversational Primes: How a Magic Trick Unconsciously Influences Hard Choices." *Proceedings of the National Academy of Sciences of the United States of America* 117, no. 30 (July 2020): 17675–17679. https://doi.org/10.1073 /pnas.2000682117

31. Wind, A. *Before We Begin.* Vanishing Incorporated, 2021; Strivings, M. *Before the Curtain Rises: A Treatise on Preshow Work.* Vol. 1. Mark D. Strivings, 2000.

32. "Inside the Capitol Riot: An Exclusive Video Investigation." *New York Times*, June 30, 2021. https://www.nytimes.com/2021/06/30/us/jan-6-capitol-attack-takeaways.html

Chapter 7

1. Harris, G., B. Carey, and J. Roberts. "Psychiatrists, Children and Drug Industry's Role." *New York Times*, May 10, 2007. https://www.nytimes.com/2007/05/10/health/10psyche.html

2. Khan, A., J. Faucett, P. Lichtenberg, I. Kirsch, and W. A. Brown. "A Systematic Review of Comparative Efficacy of Treatments and Controls for Depression." *PLOS One* 7, no. 7 (July 2012): e41778. https://doi.org:10.1371/journal.pone.0041778

3. Khan, A., J. Faucett, P. Lichtenberg, I. Kirsch, and W. A. Brown. "A Systematic Review of Comparative Efficacy of Treatments and Controls for Depression." *PLOS One* 7, no. 7 (July 2012): e41778. https://doi.org:10.1371/journal.pone.0041778; Zou, L., et al. "Effects of Meditative Movements on Major Depressive Disorder: A Systematic Review and Meta-Analysis of Randomized Controlled Trials." *Journal of Clinical Medicine* 7, no. 8 (August 2018): 195; Macías-Cortés, E. d. C., L. Llanes-González, L. Aguilar-Faisal, and J. Asbun-Bojalil. "Individualized Homeopathic Treatment and Fluoxetine for Moderate to Severe Depression in Peri- and Postmenopausal Women (HOMDEP-MENOP Study): A Randomized, Double-Dummy, Double-Blind, Placebo-Controlled Trial." *PLOS One* 10, no. 3 (March 2015): e0118440. https://doi.org:10.1371/journal .pone.0118440; Appleton, K. M., H. M. Sallis, R. Perry, A. R. Ness, and R. Churchill. "Omega-3 Fatty Acids for Depression in Adults." *Cochrane Database of Systematic Reviews* 11, (November 2015): CD004692. https://doi.org:10.1002/14651858.CD004692.pub4

4. Khan, A., J. Faucett, P. Lichtenberg, I. Kirsch, and W. A. Brown. "A Systematic Review of Comparative Efficacy of Treatments and Controls for Depression." *PLOS One* 7, no. 7 (July 2012): e41778. https://doi.org:10.1371/journal.pone.0041778

5. Raz, A. "Perspectives on the Efficacy of Antidepressants for Child and Adolescent Depression." *PLOS Med* 3, no. 1 (January 2006): e9. https://doi.org:10.1371/journal.pmed.0030009

6. Carey, B. "Panel Wants Broader Antidepressant Labeling." *New York Times*, December 14, 2006.

7. Kirsch, I., and G. Sapirstein. "Listening to Prozac but Hearing Placebo: A Meta-Analysis of Antidepressant Medications." *Prevention & Treatment* 1, no. 2 (1998).

8. Hamilton, M. "A Rating Scale for Depression." *Journal of Neurology, Neurosurgery, and Psychiatry* 23, no. 1 (February 1960): 56–62.

9. Amrhein, V., S. Greenland, and B. McShane. "Scientists Rise Up Against Statistical Significance." *Nature* 567, no. 7748 (March 2019): 305–307.

10. Ranganathan, P., C. Pramesh, and M. Buyse. "Common Pitfalls in Statistical Analysis: Clinical Versus Statistical Significance." *Perspectives in Clinical Research* 6, no. 3 (July–September 2015): 169–170.

11. Morrison, D. E., and R. E. Henkel. *The Significance Test Controversy: A Reader*. Transaction Publishers, 2006.

12. Sterne, J. A. C., and G. D. Smith. "Sifting the Evidence—What's Wrong with Significance Tests?" *Physical Therapy & Rehabilitation Journal* 81, no. 8 (August 2001): 1464–1469. https://doi .org:10.1093/ptj/81.8.1464

13. Kirk, R. E. "Practical Significance: A Concept Whose Time Has Come." *Educational and Psychological Measurement* 56, no. 5 (1996): 746–759. https://doi.org:10.1177/0013164496 056005002

14. Ogles, B. M., K. M. Lunnen, and K. Bonesteel. "Clinical Significance: History, Application, and Current Practice." *Clinical Psychology Review* 21, no. 3 (April 2001): 421–446. https://doi .org/10.1016/S0272-7358(99)00058-6; Millis, S. R. "Emerging Standards in Statistical Practice: Implications for Clinical Trials in Rehabilitation Medicine." *American Journal of Physical Medicine & Rehabilitation* 82, no. 10 Suppl. (October 2003): S32–37. https://doi.org:10.1097/01 .Phm.0000087007.19214.32

15. LeFort, S. M. "The Statistical Versus Clinical Significance Debate." *Journal of Nursing Scholarship* 25, no. 1 (1993): 57–62. https://doi.org:10.1111/j.1547-5069.1993.tb00754.x

16. Fethney, J. "Statistical and Clinical Significance, and How to Use Confidence Intervals to Help Interpret Both." *Australian Critical Care* 23, no. 2 (May 2010): 93–97. https://doi.org:10.1016/j .aucc.2010.03.001

17. Williamson, D. A., G. A. Bray, and D. H. Ryan. "Is 5% Weight Loss a Satisfactory Criterion to Define Clinically Significant Weight Loss?" *Obesity* 23, no. 12 (December 2015): 2319–2320. https://doi.org:10.1002/oby.21358

18. Ross, R., and A. J. Bradshaw. "The Future of Obesity Reduction: Beyond Weight Loss." *Nature Reviews Endocrinology* 5, no. 6 (June 2009): 319–325. https://doi.org:10.1038/nrendo.2009.78

19. Laaksonen, D. E., et al. "Physical Activity in the Prevention of Type 2 Diabetes: The Finnish Diabetes Prevention Study." *Diabetes* 54, no. 1 (January 2005): 158–165. https://doi.org:10.2337 /diabetes.54.1.158

20. Hamman, R. F., et al. "Effect of Weight Loss with Lifestyle Intervention on Risk of Diabetes." *Diabetes Care* 29, no. 9 (September 2006): 2102–2107. https://doi.org:10.2337/dc06-0560

21. Ross, R. "Is Setting a Criterion for 'Clinically Significant Weight Loss' Necessary?" *Obesity* 24, no. 4 (April 2016): 791.

22. Stone, M. B., et al. "Response to Acute Monotherapy for Major Depressive Disorder in Random-ized, Placebo Controlled Trials Submitted to the US Food and Drug Administration: Individual Participant Data Analysis." *BMJ* 378, (August 2022): e067606. https://doi.org:10.1136/bmj -2021-067606; Moncrieff, J., and I. Kirsch. "Empirically Derived Criteria Cast Doubt on the Clinical Significance of Antidepressant-Placebo Differences." *Contemporary Clinical Trials* 43, (July 2015): 60–62. https://doi.org:10.1016/j.cct.2015.05.005

23. Stone, M. B., et al. "Response to Acute Monotherapy for Major Depressive Disorder in Random-ized, Placebo Controlled Trials Submitted to the US Food and Drug Administration: Individual Participant Data Analysis." *BMJ* 378, (August 2022): e067606. https://doi.org:10.1136/bmj -2021-067606

24. Even, C., E. Siobud-Dorocant, and R. M. Dardennes. "Critical Approach to Antidepressant Trials. Blindness Protection Is Necessary, Feasible and Measurable." *British Journal of Psychiatry* 177, (July 2000): 47–51. https://doi.org:10.1192/bjp.177.1.47

25. Leyburn, P. "A Critical Look at Antidepressant Drug Trials." *Lancet* 2, no. 7526 (November 1967): 1135–1138. https://doi.org:10.1016/s0140-6736(67)90635-6

26. Khan, A., J. Faucett, P. Lichtenberg, I. Kirsch, and W. A. Brown. "A Systematic Review of Com-parative Efficacy of Treatments and Controls for Depression." *PLOS One* 7, no. 7 (July 2012): e41778. https://doi.org:10.1371/journal.pone.0041778; Greenberg, R. P., R. F. Bornstein, M. D. Greenberg, and S. Fisher. "A Meta-Analysis of Antidepressant Outcome Under 'Blinder' Condi-tions." *Journal of Consulting and Clinical Psychology* 60, no. 5 (October 1992): 664–669; discussion 670–677. https://doi.org:10.1037//0022-006x.60.5.664

27. Holper, L., and M. P. Hengartner. "Comparative Efficacy of Placebos in Short-Term Antidepres-sant Trials for Major Depression: A Secondary Meta-Analysis of Placebo-Controlled Trials." *BMC Psychiatry* 20, no. 437 (September 2020). https://doi.org:10.1186/s12888-020-02839-y

28. Frank, R., J. Avorn, and A. Kesselheim. "What Do High Drug Prices Buy Us?" *Health Affairs Blog,* April 29, 2020.

29. Espay, A. J., et al. "Placebo Effect of Medication Cost in Parkinson Disease: A Randomized Double-Blind Study." *Neurology* 84, no. 8 (February 2015): 794–802; LeWitt, P. A., and S. Kim. "The Pharmacodynamics of Placebo: Expectation Effects of Price as a Proxy for Efficacy." *Neurology* 84, no. 8 (February 2015): 766–777.

30. Raz, A., and C. S. Harris. *Placebo Talks: Modern Perspectives on Placebos in Society.* Oxford: Oxford University Press, 2016.

31. Alexander, S., "SSRIs: Much More Than You Wanted to Know." *Slate Star Codex,* July 7, 2014.

32. Busner, J., and S. D. Targum. "The Clinical Global Impressions Scale: Applying a Research Tool in Clinical Practice." *Psychiatry* 4, no. 7 (July 2007): 28–37.

33. Leucht, S., et al. "What Does the HAMD Mean?" *Journal of Affective Disorders* 148, no. 2–3 (June 2013): 243–248. https://doi.org:10.1016/j.jad.2012.12.001

34. Thase, M. E. "Antidepressant Effects: The Suit May Be Small, but the Fabric Is Real." *Prevention & Treatment* 5, no. 1 (2002); Hollon, S. D., R. J. DeRubeis, R. C. Shelton, and B. Weiss. "The Emperor's New Drugs: Effect Size and Moderation Effects." *Prevention & Treatment* 5, no. 1 (2002).

35. Kirsch, I., et al. "Initial Severity and Antidepressant Benefits: A Meta-Analysis of Data Submitted to the Food and Drug Administration." *PLOS Medicine* 5, no. 2 (February 2008): e45.

36. Konishi, T., et al. "Estimation of Prevalence of Major Depression in Patients with Subacute Myelo-Optico-Neuropathy." *European Neuropsychopharmacology* (2007): S403; Lauritzen, L., et al. "Relapse Prevention by Means of Paroxetine in ECT-Treated Patients with Major Depression: A Comparison with Imipramine and Placebo in Medium-Term Continuation Therapy." *Acta Psychiatrica Scandinavica* 94, no. 4 (October 1996): 241–251; National Collaborating Centre

REFERENCES

for Mental Health. *Depression: The Treatment and Management of Depression in Adults* Updated Edition. British Psychological Society, 2010.

37. Vöhringer, P. A., and S. N. Ghaemi. "Solving the Antidepressant Efficacy Question: Effect Sizes in Major Depressive Disorder." *Clinical Therapeutics* 33, no. 12 (December 2011): B49–B61.

38. Kessler, R. C., et al. "The Epidemiology of Major Depressive Disorder: Results from the National Comorbidity Survey Replication (NCS-R)." *JAMA* 289, no. 23 (June 2003): 3095–3105.

39. Fountoulakis, K. N., and H. J. Möller. "Efficacy of Antidepressants: A Re-Analysis and Re-Interpretation of the Kirsch Data." *International Journal of Neuropsychopharmacology* 14, no. 3 (April 2011): 405–412; Fournier, J. C., et al. "Antidepressant Drug Effects and Depression Severity: A Patient-Level Meta-Analysis." *JAMA* 303, no. 1 (January 2010): 47–53; The National Institute for Health and Care Excellence. *Depression: Management of Depression in Primary and Secondary Care*. NICE Guidance, December 2004; Turner, E. H., A. M. Matthews, E. Linardatos, R. A. Tell, and R. Rosenthal. "Selective Publication of Antidepressant Trials and Its Influence on Apparent Efficacy." *New England Journal of Medicine* 358, no. 3 (January 2008): 252–260.

40. Stone, M., S. Kalaria, K. Richardville, and B. Miller. "Components and Trends in Treatment Effects in Randomized Placebo-Controlled Trials in Major Depressive Disorder from 1979–2016." *American Society of Clinical Psychopharmacology* (May 2018); Stone, M. B., et al. "Response to Acute Monotherapy for Major Depressive Disorder in Randomized, Placebo Controlled Trials Submitted to the US Food and Drug Administration: Individual Participant Data Analysis." *BMJ* 378, (August 2022): e067606. https://doi.org/10.1136/bmj-2021-067606

41. Sugarman, M. A., A. M. Loree, B. B. Baltes, E. R. Grekin, and I. Kirsch. "The Efficacy of Paroxetine and Placebo in Treating Anxiety and Depression: A Meta-Analysis of Change on the Hamilton Rating Scales." *PLOS One* 9, no. 8 (August 2014): e106337; Sugarman, M. A., I. Kirsch, and J. D. Huppert. "Obsessive-Compulsive Disorder Has a Reduced Placebo (and Antidepressant) Response Compared to Other Anxiety Disorders: A Meta-Analysis." *Journal of Affective Disorders* 218, (August 2017): 217–226.

42. Vargas, C. *Comment in response to Kirsch, et al.* 2008.

43. Werner, R. *Comment in response to Kirsch, et al.* 2008.

44. Honigfeld, G. "Non-Specific Factors in Treatment. I. Review of Placebo Reactions and Placebo Reactors." *Diseases of the Nervous System* 25, (March 1964): 145–156.

45. Inhalatorium: A Collection of Inhalers and Asthma Therapies. 2023. http://www.inhalatorium.com/the-collection/asthma-cigarettes.

46. Fels, A. "Should We All Take a Bit of Lithium?" *New York Times*, September 13, 2014.

47. Hamblin, J. "Why We Took Cocaine Out of Soda." *The Atlantic*, January 31, 2013.

48. Kirsch, I. *The Emperor's New Drugs: Exploding the Antidepressant Myth*. Basic Books, 2010.

49. Moseley, J. B., et al. "A Controlled Trial of Arthroscopic Surgery for Osteoarthritis of the Knee." *New England Journal of Medicine* 347, no. 2 (July 2002): 81–88. https://doi.org/10.1056/NEJMoa013259

50. Thorlund, J. B., C. B. Juhl, E. M. Roos, and L. S. Lohmander. "Arthroscopic Surgery for Degenerative Knee: Systematic Review and Meta-Analysis of Benefits and Harms." *BMJ* 350, (June 2015): h2747. https://doi.org/10.1136/bmj.h2747

51. Thorlund, J. B., C. B. Juhl, E. M. Roos, and L. S. Lohmander. "Arthroscopic Surgery for Degenerative Knee: Systematic Review and Meta-Analysis of Benefits and Harms." *BMJ* 350, (June 2015): h2747. https://doi.org/10.1136/bmj.h2747

52. Siemieniuk, R. A. C., et al. "Arthroscopic Surgery for Degenerative Knee Arthritis and Meniscal Tears: A Clinical Practice Guideline." *BMJ* 357, (May 2017): j1982. https://doi.org/10.1136/bmj.j1982

53. Miller, F. G. "The Enduring Legacy of Sham-Controlled Trials of Internal Mammary Artery Ligation." *Progress in Cardiovascular Diseases* 55, no. 3 (November–December 2012): 246–250. https://doi.org:10.1016/j.pcad.2012.09.002

54. Ratcliff, J. "New Surgery for Ailing Hearts." *Reader's Digest* 71, (1957): 70–73.

55. Dimond, E. G., C. F. Kittle, and J. E. Crockett. "Comparison of Internal Mammary Artery Ligation and Sham Operation for Angina Pectoris." *American Journal of Cardiology* 5, (April 1960): 483–486. https://doi.org/10.1016/0002-9149(60)90105-3; Cobb, L. A., G. I. Thomas, D. H. Dillard, K. A. Merendino, and R. A. Bruce. "An Evaluation of Internal-Mammary-Artery Ligation by a Double-Blind Technic." *New England Journal of Medicine* 260, no. 22 (May 1959): 1115–1118. https://doi.org:10.1056/nejm195905282602204

56. Raz, A., and C. S. Harris. *Placebo Talks: Modern Perspectives on Placebos in Society.* Oxford: Oxford University Press, 2016.

57. Frakt, A., and A. E. Carroll. "How to Measure a Medical Treatment's Potential for Harm." *New York Times*, February 2, 2015.

58. Shalhoub, J., et al. "Compression Stockings in Addition to Low-Molecular-Weight Heparin to Prevent Venous Thromboembolism in Surgical Inpatients Requiring Pharmacoprophylaxis: The GAPS Non-Inferiority RCT." *Health Technology Assessment* 24, no. 69 (December 2020): 1–80. https://doi.org:10.3310/hta24690

59. Amaragiri, S. V., and T. A. Lees. "Elastic Compression Stockings for Prevention of Deep Vein Thrombosis." *Cochrane Database Systematic Review* 7, (July 2000). https://doi.org/10.1002/14651858.CD0011484

60. Carroll, A. E., and A. Frakt. "How to Measure a Medical Treatment's Potential for Harm." *New York Times*, February 2, 2015.

61. Chong, C. A. K. Y., et al. "An Unadjusted NNT Was a Moderately Good Predictor of Health Benefit." *Journal of Clinical Epidemiology* 59, no. 3 (March 2006): 224–233. https://doi.org/10.1016/j.jclinepi.2005.08.005

62. McAlister, F. A. "The 'Number Needed to Treat' Turns 20—and Continues to Be Used and Misused." *Canadian Medical Association Journal* 179, no. 6 (September 2008): 549–553. https://doi.org/10.1503/cmaj.080484

63. Maher, C. G., A. C. Traeger, C. A. Shaheed, and M. O'Keeffe. "Placebos in Clinical Care: A Suggestion Beyond the Evidence." *Medical Journal of Australia* 215, no. 6 (September 2021): 252–253. https://doi.org/10.5694/mja2.51230

64. Sidis, B. "The Psychology of Suggestion." *Science* 8, (1898): 162–163. https://doi.org/10.1126/science.8.188.162; Parris, B. A., N. Hasshim, and Z. Dienes. "Look into My Eyes: Pupillometry Reveals That a Post-Hypnotic Suggestion for Word Blindness Reduces Stroop Interference by Marshalling Greater Effortful Control." *European Journal of Neuroscience* 53, no. 8 (April 2021): 2819–2834. https://doi.org/10.1111/ejn.15105; Kihlstrom, J. F. "Placebo: Feeling Better, Getting Better, and the Problems of Mind and Body." *McGill Journal of Medicine* 11, no. 2 (November 2008): 212–214.

65. Raz, A., and M. Lifshitz. *Hypnosis and Meditation: Towards an Integrative Science of Conscious Planes.* Oxford: Oxford University Press, 2016; Terhune, D. B., A. Cleeremans, A. Raz, and S. J. Lynn. "Hypnosis and Top-Down Regulation of Consciousness." *Neuroscience and Biobehavioral Reviews* 81, Pt. A (October 2017): 59–74. https://doi.org:10.1016/j.neubiorev.2017.02.002; Acunzo, D. J., D. A. Oakley, and D. B. Terhune. "The Neurochemistry of Hypnotic Suggestion." *American Journal of Clinical Hypnosis* 63, no. 4 (April 2021): 355–371. https://doi.org/10.1080/00029157.2020.1865869

66. Raz, A., and D. Guindi. "Placebos and Medical Education." *McGill Journal of Medicine* 11, no. 2 (November 2008); Raz, A., E. Raikhel, and R. D. Anbar. "Placebos in Medicine: Knowledge, Beliefs, and Patterns of Use." *McGill Journal of Medicine* 11, no. 2 (November 2008): 206–211;

Bates, G. "The Survival of Suggestion: Charles Lloyd Tuckey and British Medical Hypnotism (1888–1914)." PhD thesis, University of London, 2021.

67. Lilienfeld, S. O., and H. Arkowitz. "Can Positive Thinking Be Negative?" *Scientific American* 22, no. 2 (May 2011). https://www.scientificamerican.com/article/can-positive-thinking-be-negative

68. Kellerman, G. R. *Tomorrowmind: Thriving at Work—Now and in an Uncertain Future*. Atria Books, 2023.

69. Langer, E. *The Mindful Body: Thinking Our Way to Chronic Health*. Ballantine Books, 2023.

70. Seligman, M. E. P., and M. Csikszentmihalyi. "Positive Psychology: An Introduction." *American Psychologist* 55, no. 1 (2000): 5–14. https://doi.org/10.1037/0003-066X.55.1.5; Held, B. S. "The Negative Side of Positive Psychology." *Journal of Humanistic Psychology* 44, no. 1 (January 2004): 9–46. https://doi.org/10.1177/0022167803259645; Ehrenreich, B. *Bright-Sided: How the Relentless Promotion of Positive Thinking Has Undermined America*. Metropolitan Books, 2009.

71. Tsur, N., R. Defrin, C. S. Haller, K. Bercovitz, and E. J. Langer. "The Effect of Mindful Attention Training for Pain Modulation Capacity: Exploring the Mindfulness-Pain Link." *Journal of Clinical Psychology* 77, no. 4 (April 2020): 896–909. https://doi.org/10.1002/jclp.23063; Pagnini, F., et al. "Ageing as a Mindset: A Study Protocol to Rejuvenate Older Adults with a Counterclockwise Psychological Intervention." *MBL Open* 9, no. 7 (July 2019). https://doi .org/10.1136/bmjopen-2019-030411; Demers, M., et al. "Beta-Testing of an Online Mindfulness Program Designed for Stroke Survivors and Their Caregivers During a Pandemic." *Archives of Physical Medicine and Rehabilitation* 102, no. 10 (October 2021): 37. https://doi.org/10.1016/j .apmr.2021.07.569

72. Langer, E. J. *Counterclockwise: Mindful Health and the Power of Possibility*. Ballantine Books, 2009; Butler, R. N., W. H. Thomas, E. J. Langer, and T. Roszak. *Longevity Rules: How to Age Well into the Future*. Eskaton, 2010; Langer, E. J. *The Mindful Body: Creating Chronic Health in an Age of Possibility*. Ballantine Group, 2023.

73. Tversky, A., and D. Kahneman in *Multiple Criteria Decision Making and Risk Analysis Using Microcomputers*, 81–126. Springer, 1989.

74. Steinert, T. "Chance of Response to an Antidepressant: What Should We Say to the Patient?" *World Psychiatry* 17, no. 1 (February 2018): 114.

75. Beauchamp, T. L., and J. F. Childress. *Principles of Biomedical Ethics*. Oxford: Oxford University Press, 2001.

76. Cuijpers, P., et al. "The Effects of Psychotherapy for Adult Depression on Suicidality and Hopelessness: A Systematic Review and Meta-Analysis." *Journal of Affective Disorders* 144, no. 3 (January 2013): 183–190.

77. Khan, A., K. Fahl Mar, J. Faucett, S. Khan Schilling, and W. A. Brown. "Has the Rising Placebo Response Impacted Antidepressant Clinical Trial Outcome? Data from the US Food and Drug Administration 1987–2013." *World Psychiatry* 16, no. 2 (June 2017): 181–192.

78. Caspi, A., et al. "Longitudinal Assessment of Mental Health Disorders and Comorbidities Across 4 Decades Among Participants in the Dunedin Birth Cohort Study." *JAMA Network Open* 3, no. 4 (April 2020): e203221–e203221. https://doi.org:10.1001/jamanetworkopen.2020.3221

79. Almohammed, O. A., et al. "Antidepressants and Health-Related Quality of Life (HRQoL) for Patients with Depression: Analysis of the Medical Expenditure Panel Survey from the United States." *PLOS One* 17, no. 4 (April 2022): e0265928. https://doi.org:10.1371/journal .pone.0265928

80. Stone, M. B., et al. "Response to Acute Monotherapy for Major Depressive Disorder in Randomized, Placebo Controlled Trials Submitted to the US Food and Drug Administration: Individual Participant Data Analysis." *BMJ* 378, (August 2022): e067606. https://doi.org:10.1136/bmj -2021-067606

81. Horowitz, M., and M. Wilcock. "Newer Generation Antidepressants and Withdrawal Effects: Reconsidering the Role of Antidepressants and Helping Patients to Stop." *Drug and Therapeutics Bulletin* 60, no. 1 (January 2022): 7–12. https://doi.org:10.1136/dtb.2020.000080

82. Read, J., C. Cartwright, and K. Gibson. "Adverse Emotional and Interpersonal Effects Reported by 1829 New Zealanders While Taking Antidepressants." *Psychiatry Research* 216, no. 1 (April 2014): 67–73. https://doi.org:10.1016/j.psychres.2014.01.042

83. Moncrieff, J. "Against the Stream: Antidepressants Are Not Antidepressants—an Alternative Approach to Drug Action and Implications for the Use of Antidepressants." *BJPsych Bulletin* 42, no. 1 (February 2018): 42–44. https://doi.org:10.1192/bjb.2017.11

84. McCain, J. A. "Antidepressants and Suicide in Adolescents and Adults: A Public Health Experiment with Unintended Consequences?" *Pharmacy and Therapeutics* 34, no. 7 (July 2009): 355–367, 378.

85. Hengartner, M. P., and M. Plöderl. "Newer-Generation Antidepressants and Suicide Risk in Randomized Controlled Trials: A Re-Analysis of the FDA Database." *Psychotherapy and Psychosomatics* 88, no. 4 (June 2019): 247–248. https://doi.org:10.1159/000501215

86. Raz, A., and R. Michels. "Contextualizing Specificity: Specific and Non-Specific Effects of Treatment." *American Journal of Clinical Hypnosis* 50, no. 2 (October 2007): 177–182.

87. Raz, A., et al. "Placebos in Clinical Practice: Comparing Attitudes, Beliefs, and Patterns of Use Between Academic Psychiatrists and Nonpsychiatrists." *Canadian Journal of Psychiatry* 56, no. 4 (April 2011): 198–208.

88. Angell, M. "The Illusions of Psychiatry." *New York Review of Books*, July 14, 2011, 20–22.

89. Haas, J. W., et al. "Frequency of Adverse Events in the Placebo Arms of COVID-19 Vaccine Trials: A Systematic Review and Meta-Analysis." *JAMA Network Open* 5, no. 1 (January 2022) e2143955; Staff article. "Two-Thirds of Corona Jab Reactions Caused by Placebo Effect—Study." *Jerusalem Post*, January 20, 2022.

90. Kuhn, T. *The Structure of Scientific Revolutions*. Princeton University Press, 2021.

91. Kuhn, T. *The Structure of Scientific Revolutions*, 139, 159. Princeton University Press, 2021.

92. Kuhn, T. *The Structure of Scientific Revolutions*, 37, 144. Princeton University Press, 2021.

93. Kuhn, T. *The Structure of Scientific Revolutions*, 40, 41, 52, 175. Princeton University Press, 2021.

94. Kuhn, T. *The Structure of Scientific Revolutions*, 109, 111. Princeton University Press, 2021.

95. Dolby, R. "Philosophy of Science—Criticism and the Growth of Knowledge. Proceedings of the International Colloquium in the Philosophy of Science, London 1965, Volume 4," review of same, edited by I. Lakatos and A. Musgrave. *British Journal for the History of Science* 5, no. 4 (December 1971).

96. Lakatos, I., and A. Musgrave, eds. *Criticism and the Growth of Knowledge. Vol. 4 of Proceedings of the International Colloquium in the Philosophy of Science, London, 1965*. Cambridge: Cambridge University Press, 1970.

97. Planck, M. *Scientific Autobiography: And Other Papers*. Open Road Media, 2014.

98. Angell, M. "The Illusions of Psychiatry." *New York Review of Books*, July 14, 2011, 20–22.

99. Bonin, R., producer. *Sixty Minutes*. "Treating Depression." CBS Media Ventures, February 19, 2012.

100. Begley, S. "Why Antidepressants Are No Better Than Placebos." *Newsweek*, January, 29 2010.

101. Vahdat, S., L. Hamzehgardeshi, S. Hessam, and Z. Hamzehgardeshi. "Patient Involvement in Health Care Decision Making: A Review." *Iranian Red Crescent Medical Journal* 16, no. 1 (January 2014).

REFERENCES

102. Khan, A., J. Faucett, P. Lichtenberg, I. Kirsch, and W. A. Brown. "A Systematic Review of Comparative Efficacy of Treatments and Controls for Depression." *PLOS One* 7, no. 7 (July 2012): e41778. https://doi.org:10.1371/journal.pone.0041778; Zou, L., et al. "Effects of Meditative Movements on Major Depressive Disorder: A Systematic Review and Meta-Analysis of Randomized Controlled Trials." *Journal of Clinical Medicine* 7, no. 8 (August 2018): 195; Macías-Cortés, E. d. C., L. Llanes-González, L. Aguilar-Faisal, and J. Asbun-Bojalil. "Individualized Homeopathic Treatment and Fluoxetine for Moderate to Severe Depression in Peri- and Postmenopausal Women (HOMDEP-MENOP Study): A Randomized, Double-Dummy, Double-Blind, Placebo-Controlled Trial." *PLOS One* 10, no. 3 (March 2015): e0118440. https://doi.org:10.1371/journal.pone.0118440; Appleton, K. M., H. M. Sallis, R. Perry, A. R. Ness, and R. Churchill. "Omega-3 Fatty Acids for Depression in Adults." *Cochrane Database of Systematic Reviews* 11, (November 2015): CD004692. https://doi.org:10.1002/14651858.CD004692.pub4

103. Hollon, S. D., P. W. Andrews, D. R. Singla, M. M. Maslej, and B. H. Mulsant. "Evolutionary Theory and the Treatment of Depression: It Is All About the Squids and the Sea Bass." *Behaviour Research and Therapy* 143, (August 2021): 103849. https://doi.org:10.1016/j.brat.2021.103849

104. Cuijpers, P., et al. "Does Cognitive Behaviour Therapy Have an Enduring Effect That Is Superior to Keeping Patients on Continuation Pharmacotherapy? A Meta-Analysis." *BMJ Open* 3, (2013).

105. Sander, L. B., et al. "Effects of Digital Cognitive Behavioral Therapy for Depression on Suicidal Thoughts and Behavior: Protocol for a Systematic Review and Meta-Analysis of Individual Participant Data." *PLOS One* 18, no. 6 (June 2023): e0285622. https://doi.org:10.1371/journal.pone.0285622

106. Cuijpers, P., et al. "Cognitive Behavior Therapy vs. Control Conditions, Other Psychotherapies, Pharmacotherapies and Combined Treatment for Depression: A Comprehensive Meta-Analysis Including 409 Trials with 52,702 Patients." *World Psychiatry* 22, no. 1 (February 2023): 105–115. https://doi.org:10.1002/wps.21069

107. Babyak, M., et al. "Exercise Treatment for Major Depression: Maintenance of Therapeutic Benefit at 10 Months." *Psychosomatic Medicine* 62, no. 5 (September–October 2000): 633–638. https://doi.org:10.1097/00006842-200009000-00006

108. Moncrieff, J., et al. "The Serotonin Theory of Depression: A Systematic Umbrella Review of the Evidence." *Molecular Psychiatry* 28, (July 2022): 3243–3256. https://doi.org:10.1038/s41380-022-01661-0

109. Horowitz, M. A., K. Munkholm, and J. Moncrieff. "Unbalanced Appraisal of Psychosocial Versus Antipsychotic Literature." *Lancet Psychiatry* 9, no. 7 (July 2022): 540–541. https://doi.org:10.1016/S2215-0366(22)00153-5; Moncrieff, J., and M. Horowitz. "Author Reply." *British Journal of Psychiatry* 221, no. 2 (August 2022): 497–498. https://doi.org:10.1192/bjp.2022.59; Moncrieff, J., and M. Horowitz. "Why It Is Important to Discuss What Antidepressants Do." *BMJ* 379, (October 2022). https://doi.org:10.1136/bmj.o2350

110. Kihlstrom, J. F. "Placebo: Feeling Better, Getting Better, and the Problems of Mind and Body." *McGill Journal of Medicine* 11, no. 2 (November 2008): 212–214.

111. Guerra-Doce, E., et al. "Direct Evidence of the Use of Multiple Drugs in Bronze Age Menorca (Western Mediterranean) from Human Hair Analysis." *Scientific Reports* 13, no. 1 (April 2023): 4782. https://doi.org:10.1038/s41598-023-31064-2

112. Dyck, E., and C. Elcock. *Expanding Mindscapes: A Global History of Psychedelics.* MIT Press, 2023.

113. Winkelman, M., and B. Sessa. *Advances in Psychedelic Medicine: State-of-the-Art Therapeutic Applications.* Santa Barbara: Praeger, 2019.

114. Borrell, B. "The Psychedelic Evangelist." *New York Times*, March 21, 2024. https://www.nytimes.com/2024/03/21/health/psychedelics-roland-griffiths-johns-hopkins.html

115. Raz, A. "Microdosing with Classical Psychedelics: Research Considerations and Practical Trajectories." Raz Lab. 2021. https://razlab.org/wp-content/uploads/2021/11/CleanBlind_R3_Microdosing -with-Classical-Psychedelics-Research-Trajectories-and-Practical-Considerations-copy.pdf

116. Dolder, P. C., Y. Schmid, F. Müller, S. Borgwardt, and M. E. Liechti. "LSD Acutely Impairs Fear Recognition and Enhances Emotional Empathy and Sociality." *Neuropsychopharmacology* 41, no. 11 (October 2016): 2638–2646.

117. Carhart-Harris, R. L., et al. "LSD Enhances Suggestibility in Healthy Volunteers." *Psychopharmacology* 232, no. 4 (February 2015): 785–794.

118. Hartogsohn, I. *American Trip: Set, Setting, and the Psychedelic Experience in the Twentieth Century.* MIT Press, 2020.

119. Hartogsohn, I. "Set and Setting, Psychedelics and the Placebo Response: An Extra-Pharmacological Perspective on Psychopharmacology." *Journal of Psychopharmacology* 30, no. 12 (December 2016): 1259–1267.

120. Petri, G., et al. "Homological Scaffolds of Brain Functional Networks." *Journal of the Royal Society Interface* 11, no. 101 (December 2014): 20140873.

121. Davis, A. K., et al. "Effects of Psilocybin-Assisted Therapy on Major Depressive Disorder: A Randomized Clinical Trial." *JAMA Psychiatry* 78, no. 5 (May 2021): 481–489. https://doi .org:10.1001/jamapsychiatry.2020.3285

122. Carhart-Harris, R. L., et al. "Psilocybin with Psychological Support for Treatment-Resistant Depression: An Open-Label Feasibility Study." *The Lancet Psychiatry* 3, no. 7 (July 2016): 619–627. https://doi.org:10.1016/s2215-0366(16)30065-7

123. O'Shaughnessy, D. M., and I. Berlowitz. "Amazonian Medicine and the Psychedelic Revival: Considering the 'Dieta.'" *Frontiers in Pharmacology* 12, (May 2021); Winkelman, M. J. "The Evolved Psychology of Psychedelic Set and Setting: Inferences Regarding the Roles of Shamanism and Entheogenic Ecopsychology." *Frontiers in Pharmacology* 12, (February 2021); Perkins, D., et al. "Influence of Context and Setting on the Mental Health and Wellbeing Outcomes of Ayahuasca Drinkers: Results of a Large International Survey." *Frontiers in Pharmacology* 12, (April 2021); Rush, B., et al. "Protocol for Outcome Evaluation of Ayahuasca-Assisted Addiction Treatment: The Case of Takiwasi Center." *Frontiers in Pharmacology* 12, (May 2021); Kettner, H., et al. "Psychedelic Communitas: Intersubjective Experience During Psychedelic Group Sessions Predicts Enduring Changes in Psychological Wellbeing and Social Connectedness." *Frontiers in Pharmacology* 12, (March 2021); Roseman, L., et al. "Relational Processes in Ayahuasca Groups of Palestinians and Israelis." *Frontiers in Pharmacology* 12, (May 2021); Hartogsohn, I. "Set and Setting in the Santo Daime." *Frontiers in Pharmacology* 12, (May 2021); Gonzalez, D., et al. "The Shipibo Ceremonial Use of Ayahuasca to Promote Well-Being: An Observational Study." *Frontiers in Pharmacology* 12, (May 2021).

124. Muttoni, S., M. Ardissino, and C. John. "Classical Psychedelics for the Treatment of Depression and Anxiety: A Systematic Review." *Journal of Affective Disorders* 258, (November 2019): 11–24; Erritzoe, D., et al. "Effects of Psilocybin Therapy on Personality Structure." *Acta Psychiatrica Scandinavica* 138, no. 5 (November 2018): 368–378. https://doi.org:10.1111/acps.12904; Goldhill, O. "Largest Psilocybin Trial Finds the Psychedelic Is Effective in Treating Serious Depression." *Stat,* November 9, 2021; Mertens, L. J., et al. "Therapeutic Mechanisms of Psilocybin: Changes in Amygdala and Prefrontal Functional Connectivity During Emotional Processing After Psilocybin for Treatment-Resistant Depression." *Journal of Psychopharmacology* 34, no. 2 (February 2020): 167–180; Murrough, J. W., et al. "Antidepressant Efficacy of Ketamine in Treatment-Resistant Major Depression: A Two-Site Randomized Controlled Trial." *American Journal of Psychiatry* 170, no. 10 (October 2013): 1134–1142.

125. Mithoefer, M. C., C. S. Grob, and T. D. Brewerton. "Novel Psychopharmacological Therapies for Psychiatric Disorders: Psilocybin and MDMA." *The Lancet Psychiatry* 3, no. 5 (May 2016): 481–488.

126. Borrell, B. "The Psychedelic Evangelist." *New York Times,* March 21, 2024. https://www.nytimes .com/2024/03/21/health/psychedelics-roland-griffiths-johns-hopkins.html

127. Gründer, G., et al. "Are Psychedelics Fast Acting Antidepressant Agents?" *Der Nervenarzt* 93, no. 3 (March 2022): 254–262; Dos Santos, R. G., J. E. Hallak, G. Baker, and S. Dursun. "Hallucinogenic/Psychedelic 5HT2A Receptor Agonists as Rapid Antidepressant Therapeutics: Evidence and Mechanisms of Action." *Journal of Psychopharmacology* 35, no. 4 (April 2021): 453–458; Kadriu, B., et al. "Ketamine and Serotonergic Psychedelics: Common Mechanisms Underlying the Effects of Rapid-Acting Antidepressants." *International Journal of Neuropsychopharmacology* 24, no. 1 (January 2021): 8–21.

128. Langlitz, N., and A. K. Gearin. "Psychedelic Therapy as Form of Life." *Neuroethics* 17, no. 1 (2024). https://doi.org/10.1007/s12152-024-09550-9

129. Davis, A. K., et al. "Effects of Psilocybin-Assisted Therapy on Major Depressive Disorder: A Randomized Clinical Trial." *JAMA Psychiatry* 78, no. 5 (May 2021): 481–489. https://doi.org:10.1001/jamapsychiatry.2020.3285; Carhart-Harris, R. L., et al. "Psilocybin with Psychological Support for Treatment-Resistant Depression: An Open-Label Feasibility Study." *The Lancet Psychiatry* 3, no. 7 (July 2016): 619–627. https://doi.org:10.1016/s2215-0366(16)30065-7; Carhart-Harris, R. L., et al. "Psilocybin for Treatment-Resistant Depression: fMRI-Measured Brain Mechanisms." *Scientific Reports* 7, no. 1 (October 2017): 13187. https://doi.org:10.1038/s41598-017-13282-7

130. Mitchell, J. M., et al. "MDMA-Assisted Therapy for Severe PTSD: A Randomized, Double-Blind, Placebo-Controlled Phase 3 Study." *Nature Medicine* 27, (2021): 1025–1033. https://doi.org:10.1038/s41591-021-01336-3

131. Nutt, D. J., L. A. King, L. D. Phillips, and Independent Scientific Committee on Drugs. "Drug Harms in the UK: A Multicriteria Decision Analysis." *Lancet* 376, no. 9752 (November 2010): 1558–1565. https://doi.org:10.1016/S0140-6736(10)61462-6

132. Borrell, B. "The Psychedelic Evangelist." *New York Times*, March 21, 2024. https://www.nytimes.com/2024/03/21/health/psychedelics-roland-griffiths-johns-hopkins.html

133. Pollan, M. *How to Change Your Mind: What the New Science of Psychedelics Teaches Us About Consciousness, Dying, Addiction, Depression, and Transcendence.* Penguin Books, 2019.

134. Doblin, R. "The Future of Psychedelic-Assisted Psychotherapy." *TED*, July 9, 2019. https://www.ted.com/talks/rick_doblin_the_future_of_psychedelic_assisted_psychotherapy

135. Rivlin, A., and J. Sharpe. "The Investment Surge in Psychedelics for Treatment of Mental Health Conditions." *Mental and Behavioral Health* 13, no. 40 (October 2021).

136. Compass Pathways. "Psilocybin Therapy: FDA Breakthrough Therapy Designation Received." *Compass Pathways* (October 24, 2022). https://compasspathways.com/compass-pathways-receives-fda-breakthrough-therapy-designation-for-psilocybin-therapy-for-treatment-resistant-depression/; Nichols, D. "Faculty Opinions Recommendation of Trial of Psilocybin Versus Escitalopram for Depression." Faculty Opinions—*Post-Publication Peer Review of the Biomedical Literature* (2021). https://doi.org/10.3410/f.739937798.793584706; Nayak, S. M., et al. "A Bayesian Reanalysis of a Trial of Psilocybin Versus Escitalopram for Depression." *Psychedelic Medicine* (2022). https://doi.org/10.31234/osf.io/sb5ur; Szigeti, B., et al. "Assessing Expectancy and Suggestibility in a Trial of Escitalopram v. Psilocybin for Depression." *Psychological Medicine*, (2024): 1–8. https://doi.org/10.1017/s0033291723003653; Carhart-Harris, R., et al. "Trial of Psilocybin Versus Escitalopram for Depression." *New England Journal of Medicine* 384, no. 15 (2021): 1402–1411. https://doi.org/10.1056/nejmoa2032994

137. Center, P.-F. (December 6, 2022). Disclosed: Conflicts of Interest in the Psychedelics…: Ev. The Petrie-Flom Center for Health Law Policy, Biotechnology, and Bioethics at Harvard Law School. https://petrieflom.law.harvard.edu/events/details/conflicts-interest-psychedelics-ecosystem; Mosurinjohn, S., L. Roseman, and M. Girn. "Psychedelic-Induced Mystical Experiences: An Interdisciplinary Discussion and Critique." *Frontiers in Psychiatry* 14 (2023). https://doi.org/10.3389/fpsyt.2023.1077311

138. Johnson, M. W. "Consciousness, Religion, and Gurus: Pitfalls of Psychedelic Medicine." *ACS Pharmacology & Translational Science* 4, no. 2 (2020): 578–581. https://doi.org/10.1021/acsptsci .0c00198; Strassman, R. J. "The Psychedelic Religion of Mystical Consciousness." *Journal of Psychedelic Studies* 2, no. 1 (2018): 1–4. https://doi.org/10.1556/2054.2018.003; Ko, K., G. Knight, J. J. Rucker, and A. J. Cleare. (2022). "Psychedelics, Mystical Experience, and Therapeutic Efficacy: A Systematic Review." *Frontiers in Psychiatry* 13 (2022). https://doi.org/10.3389 /fpsyt.2022.917199; van Elk, M. Erratum to "History Repeating: Guidelines to Address Common Problems in Psychedelic Science." *Therapeutic Advances in Psychopharmacology* 14 (2024). https:// doi.org/10.1177/20451253231223609; van Elk, M., and E. I. Fried. "History Repeating: A Roadmap to Address Common Problems in Psychedelic Science" *Therapeutic Advances in Psychopharmacology* (2023): 13. https://doi.org/10.31234/osf.io/ak6gx; Welker, J. "The Dangers of a Psychedelic Gospel." ChristianityToday.com. (November 15, 2023). https://www.christianitytoday .com/ct/2023/november-web-only/psychedelics-drug-trials-clergy-risks-christian-gospel.html

139. Mullard, A. "Will Psychedelics Be 'a Revolution in Psychiatry'?" *Nature Reviews Drug Discovery* 20, no. 6 (June 2021): 418–419; Leibowitz, K. A., L. C. Howe, and A. J. Crum. "Changing Mindsets About Side Effects." *BMJ Open* 11, no. 2 (February 2021): e040134; Cameron, L. P., and D. E. Olson. "The Evolution of the Psychedelic Revolution." *Neuropsychopharmacology* 47, no. 1 (January 2022): 413–414.

140. Lyubomirsky, S. "Toward a New Science of Psychedelic Social Psychology: The Effects of MDMA (Ecstasy) on Social Connection." *Perspectives on Psychological Science* 17, no. 5 (September 2022): 1234–1257. https://doi.org:10.1177/17456916211055369; Weiss, B., V. Nygart, L. M. Pommerencke, R. L. Carhart-Harris, and D. Erritzoe. "Examining Psychedelic-Induced Changes in Social Functioning and Connectedness in a Naturalistic Online Sample Using the Five-Factor Model of Personality." *Frontiers in Psychology* 12, (November 2021): 749788. https://doi.org:10.3389 /fpsyg.2021.749788

141. Timmermann, C., R. Watts, and D. Dupuis. "Towards Psychedelic Apprenticeship: Developing a Gentle Touch for the Mediation and Validation of Psychedelic-Induced Insights and Revelations." *Transcultural Psychiatry* 59, no. 5 (October 2022): 691–704. https://doi .org:10.1177/13634615221082796; Hartogsohn, I. "Modalities of the Psychedelic Experience: Microclimates of Set and Setting in Hallucinogen Research and Culture." *Transcultural Psychiatry* 59, no. 5 (October 2022): 579–591. https://doi.org:10.1177/13634615221100385; Gobbi, G., A. Inserra, K. T. Greenway, M. Lifshitz, and L. J. Kirmayer. "Psychedelic Medicine at a Crossroads: Advancing an Integrative Approach to Research and Practice." *Transcultural Psychiatry* 59, no. 5 (October 2022): 718–724. https://doi.org:10.1177/13634615221119388; Carhart-Harris, R. L., et al. "LSD Enhances Suggestibility in Healthy Volunteers." *Psychopharmacology* 232, no. 4 (2014): 785–794. https://doi.org/10.1007/s00213-014-3714-z; Dupuis, D., and S. Veissière, "Culture, Context, and Ethics in the Therapeutic Use of Hallucinogens: Psychedelics as Active Super-Placebos?" *Transcultural Psychiatry* 59, no. 5 (2022): 571–578. https://doi.org/10.1177 /13634615221131465

Chapter 8

1. Carhart-Harris, R. L., et al. "Neural Correlates of the LSD Experience Revealed by Multimodal Neuroimaging." *Proceedings of the National Academy of Sciences* 113, no. 17 (April 2016): 4853–4858.

2. Olson, D. E. "Biochemical Mechanisms Underlying Psychedelic-Induced Neuroplasticity." *Biochemistry* 61, no. 3 (January 2022): 127–136; Innocenti, G. M. "Defining Neuroplasticity." *Handbook of Clinical Neurology* 184, (2022): 3–18; Aleksandrova, L. R., and A. G. Phillips. "Neuroplasticity as a Convergent Mechanism of Ketamine and Classical Psychedelics." *Trends in Pharmacological Sciences* 42, no. 11 (November 2021): 929–942; Artin, H., S. Zisook, and D. Ramanathan. "How Do Serotonergic Psychedelics Treat Depression: The Potential Role of Neuroplasticity." *World Journal of Psychiatry* 11, no. 6 (June 2021): 201–214.

3. McMillan, R. M. "Prescribing Meaning: Hedonistic Perspectives on the Therapeutic Use of Psychedelic-Assisted Meaning Enhancement." *Journal of Medical Ethics* 47, no. 5 (November 2021): 701–705; Miceli McMillan, R., and C. Jordens. "Psychedelic-Assisted Psychotherapy for the Treatment of Major Depression: A Synthesis of Phenomenological Explanations." *Medicine, Health Care and Philosophy* 25, no. 2 (June 2022); Zeifman, R. J., et al. "Decreases in Suicidality Following Psychedelic Therapy: A Meta-Analysis of Individual Patient Data Across Clinical Trials." *Journal of Clinical Psychiatry* 83, no. 2 (January 2022).

4. Moerman, D. E., and W. B. Jonas. "Toward a Research Agenda on Placebo." *Advances in Mind-Body Medicine* 16, no. 1 (2000): 33–46. https://doi.org:10.1054/ambm.2000.0116; Moerman, D. E. *Meaning, Medicine, and the "Placebo Effect."* Cambridge University Press, 2002; Moerman, D. E., and W. B. Jonas. "Deconstructing the Placebo Effect and Finding the Meaning Response." *Annals of Internal Medicine* 136, no. 6 (March 2002): 471–476. https://doi.org:10.7326/0003 -4819-136-6-200203190-00011

5. McGovern, H., P. Leptourgos, B. Hutchinson, and P. R. Corlett. "Do Psychedelics Change Beliefs?" *Psychopharmacology* 239, no. 6 (June 2022): 1809–1821; Timmermann, C., et al. "Psychedelics Alter Metaphysical Beliefs." *Scientific Reports* 11, (November 2021).

6. Wong, A., and A. Raz. "Microdosing with Classical Psychedelics: Research Trajectories and Practical Considerations." *Transcultural Psychiatry* 59, no. 5 (October 2022): 675–690. https:// doi.org:10.1177/13634615221129115; Kaypak, A. C., and A. Raz. "Macrodosing to Microdosing with Psychedelics: Clinical, Social, and Cultural Perspectives." *Transcultural Psychiatry* 59, no. 5 (October 2022): 665–674. https://doi.org:10.1177/13634615221119386

7. Marks, M. "A Strategy for Rescheduling Psilocybin." *Scientific American*, October 11, 2021. https://www.scientificamerican.com/article/a-strategy-for-rescheduling-psilocybin

8. Marchese, D. "A Psychedelics Pioneer Takes the Ultimate Trip." *New York Times*, April 3, 2023.

9. Nour, M. M., L. Evans, D. Nutt, and R. L. Carhart-Harris. "Ego-Dissolution and Psychedelics: Validation of the Ego-Dissolution Inventory (EDI)." *Frontiers in Human Neuroscience* 10, (June 2016): 269.

10. Kaypak, A. C., and A. Raz. "Macrodosing to Microdosing with Psychedelics: Clinical, Social, and Cultural Perspectives." *Transcultural Psychiatry* 59, no. 5 (October 2022): 665–674.

11. Tart, C. T. *On Being Stoned: A Psychological Study of Marijuana Intoxication.* Big Sur Recordings, 1971.

12. Olson, J. A., L. Suissa-Rocheleau, M. Lifshitz, A. Raz, and S. P. Veissière. "Tripping on Nothing: Placebo Psychedelics and Contextual Factors." *Psychopharmacology* 237, no. 5 (May 2020): 1371–1382.

13. Asch, S. E. "Opinions and Social Pressure." *Scientific American* 193, no. 5 (November 1955): 31–35; Asch, S. E. In *Groups, Leadership and Men*, edited by H. Guetzkow. Carnegie Press, 1951; Asch, S. E., "Studies of Independence and Conformity: I. A Minority of One Against a Unanimous Majority." *Psychological Monographs: General and Applied* 70, no. 9 (1956): 1–70.

14. Olson, J. A., L. Suissa-Rocheleau, M. Lifshitz, A. Raz, and S. P. Veissière. "Tripping on Nothing: Placebo Psychedelics and Contextual Factors." *Psychopharmacology* 237, no. 5 (May 2020): 1371–1382.

15. Hendricks, P. S. "Awe: A Putative Mechanism Underlying the Effects of Classic Psychedelic-Assisted Psychotherapy." *International Review of Psychiatry* 30, no. 4 (August 2018): 331–342. https://doi.org:10.1080/09540261.2018.1474185

16. Cooper, L. "Could Group Therapy Get a Boost from Psychedelics?" *Wall Street Journal*, March 10, 2021.

17. Hartogsohn, I. "Constructing Drug Effects: A History of Set and Setting." *Drug Science, Policy and Law* 3, (January 2017).

REFERENCES

18. Olson, J. A., L. Suissa-Rocheleau, M. Lifshitz, A. Raz, and S. P. Veissière. "Tripping on Nothing: Placebo Psychedelics and Contextual Factors." *Psychopharmacology* 237, no. 5 (May 2020): 1371–1382.

19. Hartogsohn, I. "Set and Setting, Psychedelics and the Placebo Response: An Extra-Pharmacological Perspective on Psychopharmacology." *Journal of Psychopharmacology* 30, no. 12 (December 2016): 1259–1267.

20. dos Santos, R. A., and L. B. Martins. "Visions from Other Worlds: Western Esotericism, UFO Beliefs and Conspiracy Theories in New Age Ayahuasca Groups." *La Rosa di Paracelso* (2021).

21. Matthews, J. B. "Truth and Truthiness: Evidence, Experience, and Clinical Judgement in Surgery." *British Journal of Surgery* 108, no. 7 (July 2021): 742–744; Donaldson, C. "Got a Gut Feeling: Truthiness, Conspiracy and Archives in Contemporary Culture." Master's thesis, Queensland College of Art, 2021; Newman, E. J., and L. Zhang. "Truthiness." In *The Psychology of Fake News*, edited by R. Greifeneder, M. Jaffe, E. Newman, and N. Schwarz, 90–114. Routledge, 2020.

22. Lucas, T., J. Kumaratilake, and M. Henneberg. "Recently Increased Prevalence of the Human Median Artery of the Forearm: A Microevolutionary Change." *Journal of Anatomy* 237, no. 4 (October 2020): 623–631. https://doi.org:10.1111/joa.13224

23. Hochberg, Z. *Evo-Devo of Child Growth: Treatise on Child Growth and Human Evolution.* Wiley-Blackwell, 2012.

24. Friedman, T. L. "We Are Opening the Lids on Two Giant Pandora's Boxes." *New York Times*, May 2, 2023.

25. Chomsky, N. *Language and Mind.* 3rd ed. Cambridge: Cambridge University Press, 2006; Chomsky, N. *New Horizons in the Study of Language and Mind.* Cambridge: Cambridge University Press, 2000; Chomsky, N. *Language and Mind.* Harcourt, 1968.

26. Dor, D. "Communication for Collaborative Computation: Two Major Transitions in Human Evolution." *Philosophical Transactions of the Royal Society of London. Series B Biological Sciences* 378, (March 2023). https://doi.org:10.1098/rstb.2021.0404

27. Szilagyi, A., V. P. Kovacs, T. Czaran, and E. Szathmary. "Evolutionary Ecology of Language Origins Through Confrontational Scavenging." *Philosophical Transactions of the Royal Society of London. Series B Biological Sciences* 378, no. 1872 (March 2023). https://doi.org:10.1098/rstb.2021.0411

28. Rape, Abuse & Incest National Network. "The Criminal Justice System: Statistics." RAINN.org, 2021. https://www.rainn.org/statistics/criminal-justice-system

29. Lodrick, Z. "Psychological Trauma—What Every Trauma Worker Should Know." *British Journal of Psychotherapy Integration* 4, (January 2007).

30. Burrowes, N. NB Research, 2013.

31. Hardy, A., K. Young, and E. A. Holmes. "Does Trauma Memory Play a Role in the Experience of Reporting Sexual Assault During Police Interviews? An Exploratory Study." *Memory* 17, no. 3 (November 2008): 783–788. https://doi.org/10.1080/09658210903081835

32. Langer, E. J. *Counterclockwise: Mindful Health and the Power of Possibility.* 1st ed. Ballantine Books, 2009.

33. Raz, A., and S. Rabipour. *How (Not) to Train the Brain: Enhancing What's Between Your Ears with (and Without) Science.* Oxford: Oxford University Press, 2019.

34. Niau, D. *The Devils of Loudun.* Loedsak, 2011.

35. Golden, T. "The Toll Workers Illness—Was It All in Their Minds?" *New York Times*, February 1990.

36. Dominus, S. "What Happened to the Girls in Le Roy." *New York Times Magazine*, March 7, 2012.

37. Hadler, N. M. "If You Have to Prove You Are Ill, You Can't Get Well: The Object Lesson of Fibromyalgia." *Spine* 21, no. 20 (October 1996): 2397–2400.

38. Frey, J., K. J. Black, and I. A. Malaty. "TikTok Tourette's: Are We Witnessing a Rise in Functional Tic-Like Behavior Driven by Adolescent Social Media Use?" *Psychology Research and Behavior Management* 15, (December 2022): 3575–3585. https://doi.org:10.2147/PRBM.S359977

39. Forman, E. "Mysterious Hiccups Hit Essex Aggie Students." *Salem News*, January 2013. https://www.salemnews.com/news/local_news/mysterious-hiccups-hit-essex-aggie-students/article_04f5b09b-6b28-5e16-b7cc-31198d9b7cb1.html

40. Bartholomew, R. E., and R. W. Baloh. "Challenging the Diagnosis of 'Havana Syndrome' as a Novel Clinical Entity." *Journal of the Royal Society of Medicine* 113, no. 1 (January 2020): 7–11. https://doi.org:10.1177/0141076819877553

41. McCarthy, J., K. O'Donnell, L. Campbell, and D. Dooley. "Ethical Arguments for Access to Abortion Services in the Republic of Ireland: Recent Developments in the Public Discourse." *Journal of Medical Ethics* 44, no. 8 (August 2018): 513–507.

42. Taylor, M. "Women's Right to Health and Ireland's Abortion Laws." *International Journal of Gynecology and Obstetrics* 130, no. 1 (July 2015): 93–97. https://doi.org:10.1016/j.ijgo.2015.04.020

43. Murphy, G., E. F. Loftus, R. H. Grady, L. J. Levine, and C. M. Greene. "False Memories for Fake News During Ireland's Abortion Referendum." *Psychological Science* 30, no. 10 (October 2019): 1449–1459.

44. Frenda, S. J., E. D. Knowles, W. Saletan, and E. F. Loftus. "False Memories of Fabricated Political Events." *Journal of Experimental Social Psychology* 49, no. 2 (2013): 280–286. https://doi.org:10.1016/j.jesp.2012.10.013

45. Bruckman, A. *Should You Believe Wikipedia? Online Communities and the Construction of Knowledge*. Cambridge University Press, 2022.

46. Thompson, S. A. "Fox News's Settlement in the Dominion Case Is Big News, Except on Fox News." *New York Times*, April 19, 2023.

47. Tabuchi, H. "How Climate Change Deniers Rise to the Top in Google Searches." *New York Times*, December 19, 2017; Avery, D. "Facebook Ads Sharing Climate Misinformation Have Been Viewed More Than Eight MILLION Times, Despite the Firm Saying It Is Committed to Tackling the Problem." *Daily Mail*, October 2020; Allgaier, J. "Science and Environmental Communication on YouTube: Strategically Distorted Communications in Online Videos on Climate Change and Climate Engineering." *Frontiers in Communication* 4, (2019): 36. https://doi.org/10.3389/fcomm.2019.00036

48. Grinberg, N., K. Joseph, L. Friedland, B. Swire-Thompson, and D. Lazer. "Fake News on Twitter During the 2016 U.S. Presidential Election." *Science* 363, no. 6425 (January 2019): 374–378. https://doi.org:10.1126/science.aau2706

49. Vosoughi, S., D. Roy, and S. Aral. "The Spread of True and False News Online." *Science* 359, no. 6380 (March 2018): 1146–1151. https://doi.org:10.1126/science.aap9559

50. Moskalenko, S., and C. McCauley. "QAnon: Radical Opinion Versus Radical Action." *Perspectives on Terrorism* 15, no. 2 (April 2021): 142–146; Kaplan, J. A. "Conspiracy of Dunces: Good Americans vs. Cabal of Satanic Pedophiles?" *Terrorism and Political Violence* 33, no. 5 (July 2021): 917–921. https://doi.org/10.1080/09546553.2021.1932342; Bloom, M., and S. Moskalenko. *Pastels and Pedophiles: Inside the Mind of QAnon*. Redwood Press, 2021.

51. Liv, N., and D. Greenbaum. "Deep Fakes and Memory Malleability: False Memories in the Service of Fake News." *AJOB Neuroscience* 11, no. 2 (April–June 2020): 96–104. https://doi.org/10.1080/21507740.2020.1740351

52. Jay, J. *How Magicians Think: Misdirection, Deception, and Why Magic Matters.* Workman Publishing Company, 2021.

53. Appelbaum, P. S., X. Liu, A. Z. Saleh, T. S. Stroup, and M. Wall. "Deaths of People with Mental Illness During Interactions with Law Enforcement." *International Journal of Law and Psychiatry* 58, (May–June 2018): 110–116. https://doi.org:10.1016/j.ijlp.2018.03.003

54. Christian, C. W., W. G. Lane, R. Monteith, and D. Rubin. "Racial Differences in the Evaluation of Pediatric Fractures for Physical Abuse." *JAMA* 288, no. 13 (October 2002): 1603–1609. https://doi.org:10.1001/jama.288.13.1603

55. Gara, M. A., S. Minsky, T. Miskimen, S. M. Silverstein, and S. Strakowski. "A Naturalistic Study of Racial Disparities in Diagnoses at an Outpatient Behavioral Health Clinic." *Psychiatric Services* 70, no. 2 (February 2019): 130–134. https://doi.org:10.1176/appi.ps.201800223

56. Haider, A. H., et al. "Association of Unconscious Race and Social Class Bias with Vignette-Based Clinical Assessments by Medical Students." *JAMA* 306, no. 9 (September 2011): 942–951. https://doi.org/10.1001/jama.2011.1248; Green, A. R., et al. "Implicit Bias Among Physicians and Its Predictions of Thrombolysis Decisions for Black and White Patients." *Journal of General Internal Medicine* 22, no. 9 (September 2007): 1231–1238. https://doi.org/10.1007/s11606-007-0258-5

57. Dasgupta, N., and A. G. Greenwald. "On the Malleability of Automatic Attitudes: Combating Automatic Prejudice with Images of Admired and Disliked Individuals." *Journal of Personality & Social Psychology* 81, no. 5 (2001): 800–815. https://doi.org/10.1037//0022-3514.81.5.800

58. Wittenbrink, B., C. M. Judd, and B. Park. "Spontaneous Prejudice in Context: Variability in Automatically Activated Attitudes." *Journal of Personality & Social Psychology* 81, no. 5 (November 2001): 815–827. https://doi.org/10.1037//0022-35.14.81.5.815

59. Rudman, L. A., R. D. Ashmore, and M. L. Gary. "'Unlearning' Automatic Biases: The Malleability of Implicit Prejudice and Stereotypes." *Journal of Personality & Social Psychology* 81, no. 5 (2001): 856–868. https://doi.org/10.1037//0022-3514.81.5.856

60. Brehm, J. W. *A Theory of Psychological Reactance.* New York: Academic Press, 1966.

61. Naff, K. C., and E. Kellough. "Ensuring Employment Equity: Are Federal Diversity Programs Making a Difference?" *International Journal of Public Administration* 26, no. 12 (October 2003): 1307–1336; Kellough, J. E., and K. C. Naff. "Responding to a Wake-Up Call: An Examination of Federal Agency Diversity Management Programs." *Administration & Society* 36, no. 1 (2004): 62–90; Dobbin, F., K. Alexandra, and E. Kelley. "Diversity Management in Corporate America." *Contexts* 6, no. 4 (2007): 21–28; Alexandra, K., F. Dobbin, and E. Kelley. "Best Practices or Best Guesses? Assessing the Efficacy of Corporate Affirmative Action and Diversity Policies." *American Sociological Review* 71, no. 4 (2006): 589–617.

62. Onyeador, I. N., T. J. Sa-Kiera, and A. L. J. Neil. "Moving Beyond Implicit Bias Training: Policy Insights for Increasing Organizational Diversity." *Federation of Associations in Behavioral and Brain Sciences* 8, no. 1 (2021): 19–26. https://doi.org/10.1177/2372732220983840; Kollen, T. "Diversity Management: A Critical Review and Agenda for the Future." *Journal of Management Inquiry* 30, no. 3 (June 2021): 259–272. https://doi.org/10.1177/1056492619868025; Kalev, A., and F. Dobbin. "Does Diversity Training Increase Corporate Diversity? Regulation Backlash and Regulatory Accountability." Mossavar-Rahmani Center for Business & Government. July 2020. https://www.hks.harvard.edu/centers/mrcbg/programs/growthpolicy/does-diversity-training-increase-corporate-diversity-regulation

63. Stangor, C., G. B. Sechrist, and J. T. Jost. "Changing Racial Beliefs by Providing Consensus Information." *Personality and Social Psychology Bulletin* 27, no. 4 (2001): 489–496. https://doi.org/10.1177/0146167201274009

64. Maister, L., M. Slater, M. V. Sanchez-Vives, and M. Tsakiris. "Changing Bodies Changes Minds: Owning Another Body Affects Social Cognition." *Trends in Cognitive Sciences* 19, no. 1 (January

2015): 6–12. https://doi.org:10.1016/j.tics.2014.11.001; Krol, S. A., R. Thériault, J. A. Olson, A. Raz, and J. A. Bartz. "Self-Concept Clarity and the Bodily Self: Malleability Across Modalities." *Personality and Social Psychology Bulletin* 46, no. 5 (May 2019): 808–820. https://doi .org:10.1177/0146167219879126

65.   Cebolla, A., et al. "Putting Oneself in the Body of Others: A Pilot Study on the Efficacy of an Embodied Virtual Reality System to Generate Self-Compassion." *Frontiers in Psychology* 10, (July 2019). https://doi.org:10.3389/fpsyg.2019.01521

66.   Thériault, R., J. A. Olson, S. A. Krol, and A. Raz. "Body Swapping with a Black Person Boosts Empathy: Using Virtual Reality to Embody Another." *Quarterly Journal of Experimental Psychology* 74, no. 12 (December 2021): 1–18. https://doi.org:10.1177/17470218211024826

67.   Saposnik, G., et al. "Effectiveness of Virtual Reality Using Wii Gaming Technology in Stroke Rehabilitation." *Stroke* 41, no. 7 (July 2010): 1477–1484. https://doi.org/10.1161/STROKEAHA .110.584979; Moon, K. J., et al. "Physiological Response of Imagery Running with or Without an Avatar in 3D Virtual Reality: A Preliminary Study." *The Asian Journal of Kinesiology* 23, no. 3 (July 2021): 11–19. https://doi.org/10.15758/ajk.2021.23.3.11

68.   Langer, E. J. *Counterclockwise: Mindful Health and the Power of Possibility.* Ballantine Books, 2009.

69.   Raz, A., Z. R. Zephrani, H. R. Schweizer, and G. P. Marinoff. "Critique of Claims of Improved Visual Acuity After Hypnotic Suggestion." *Optometry and Vision Science* 81, no. 11 (November 2004): 872–879. https://doi.org:10.1097/01.opx.0000145032.79975.58

70.   Raz, A., G. P. Marinoff, K. Landzberg, and D. Guyton. "Substrates of Negative Accommodation." *Binocular Vision & Strabismus Quarterly* 19, no. 2 (2004): 71–74.

71.   Raz, A., G. P. Marinoff, Z. R. Zephrani, H. R. Schweizer, and M. I. Posner. "See Clearly: Suggestion, Hypnosis, Attention, and Visual Acuity." *International Journal of Clinical and Experimental Hypnosis* 52, no. 2 (April 2004): 159–187. https://doi.org:10.1076/iceh.52.2.159.28097

72.   Langer, E., M. Djikic, M. Pirson, A. Madenci, and R. Donohue. "Believing Is Seeing: Using Mindlessness (Mindfully) to Improve Visual Acuity." *Psychological Science* 21, no. 5 (May 2010): 661–666. https://doi.org:10.1177/0956797610366543

73.   Langer, E. *The Mindful Body: Thinking Our Way to Chronic Health.* Ballantine Books, 2023.

74.   Covino, N. A., D. C. Jimerson, B. E. Wolfe, D. L. Franko, and F. H. Frankel. "Hypnotizability, Dissociation, and Bulimia Nervosa." *Journal of Abnormal Psychology* 103, no. 3 (August 1994): 455–459. https://doi.org:10.1037//0021-843x.103.3.455; Barabasz, M. "Cognitive Hypnotherapy with Bulimia." *American Journal of Clinical Hypnosis* 54, no. 4 (April 2012): 353–364. https://doi .org:10.1080/00029157.2012.658122

75.   Kyaga, S., et al. "Creativity and Mental Disorder: Family Study of 300,000 People with Severe Mental Disorder." *British Journal of Psychiatry* 199, no. 5 (November 2011): 373–379. https://doi .org:10.1192/bjp.bp.110.085316; Olugbile, O., and M. P. Zachariah. "The Relationship Between Creativity and Mental Disorder in an African Setting." *Mens Sana Monographs* 9, no. 1 (January–December 2011): 225–237. https://doi.org:10.4103/0973-1229.77439; Kirov, G., and G. Miller. "Creativity and Mental Disorder." *British Journal of Psychiatry* 200, no. 4 (April 2012): 347; author reply 348. https://doi.org:10.1192/bjp.200.4.347a; Patra, B. N., and Y. P. Balhara. "Creativity and Mental Disorder." *British Journal of Psychiatry* 200, no. 4 (2012): 346; author reply 348. https:// doi.org:10.1192/bjp.200.4.346; Schmechel, D. E. "Creativity and Mental Disorder." *British Journal of Psychiatry* 200, no. 4 (2012): 347; author reply 348. https://doi.org:10.1192/bjp.200.4.347

76.   Henehan, D. "Did Shostakovich Have a Secret?" *New York Times,* July 10, 1983.

77.   Hayashibara, E., S. Savickaite, and D. Simmons. "Creativity in Neurodiversity: Towards an Inclusive Creativity Measure for Autism and ADHD." 2023.

78.   Holm-Hadulla, R. M., F. H. Hofmann, M. Sperth, and C. H. Mayer. "Creativity and Psychopathology: An Interdisciplinary View." *Psychopathology* 54, no. 1 (2021): 39–46. https://doi

.org:10.1159/000511981; Javaid, S. F., and J. P. Pandarakalam. "The Association of Creativity with Divergent and Convergent Thinking." *Psychiatria Danubina* 33, no. 2 (2021): 133–139. https://doi.org:10.24869/psyd.2021.133; Benedek, M., and A. Fink. "From Brain Images to Drawings—New Insights Informing the Creativity-Psychopathology Debate: Commentary on 'Can We Really "Read" Art to See the Changing Brain? A Review and Empirical Assessment of Clinical Case Reports and Published Artworks for Systematic Evidence of Quality and Style Changes Linked to Damage or Neurodegenerative Disease.'" *Physics of Life Reviews* 44, (March 2023): 179–183. https://doi.org:10.1016/j.plrev.2023.01.010

79.   Greengross, G., P. J. Silvia, and S. J. Crasson. "Psychotic and Autistic Traits Among Magicians and Their Relationship with Creative Beliefs." *BJPsych Open* 9, no. 6 (November 2023): e214. https://doi.org:10.1192/bjo.2023.609

Conclusion

1.   Enck, P., U. Bingel, M. Schedlowski, and W. Rief. "The Placebo Response in Medicine: Minimize, Maximize, or Personalize?" *Nature Reviews Drug Discovery* 12, no. 3 (March 2013): 191–204. https://doi.org:10.1038/nrd3923

2.   Marchant, J. "Placebo Effect Grows in U.S., Thwarting Development of Painkillers." *Scientific American*, October 7, 2015.

3.   Lynn, S. J., J. W. Rhue, and I. Kirsch. *Handbook of Clinical Hypnosis.* 2nd ed. American Psychological Association, 2010; Lynn, S. J., and I. Kirsch. *Essentials of Clinical Hypnosis: An Evidence-Based Approach.* American Psychological Association, 2006; Kirsch, I. *Clinical Hypnosis and Self-Regulation: Cognitive-Behavioral Perspectives.* American Psychological Association, 1999; Rhue, J. W., S. J. Lynn, and I. Kirsch. *Handbook of Clinical Hypnosis.* American Psychological Association, 1993.

4.   Weinberg, S. *Dreams of a Final Theory.* Vintage Books, 1994.

5.   Ben-Shaanan, T. L., et al. "Activation of the Reward System Boosts Innate and Adaptive Immunity." *Nature Medicine* 22, no. 8 (August 2016): 940–944. https://doi.org:10.1038/nm.4133

6.   Buchak, L. "Faith and Traditions." *Noûs* 57, no. 3 (September 2023): 740–759. https://doi.org/10.1111/nous.12427

7.   Zahrt, O. H., and A. J. Crum. "Perceived Physical Activity and Mortality: Evidence from Three Nationally Representative U.S. Samples." *Health Psychology* 36, no. 11 (November 2017): 1017–1025. https://doi.org:10.1037/hea0000531; Olsson, S. J. G., et al. "Association of Perceived Physical Health and Physical Fitness in Two Swedish National Samples from 1990 and 2015." *Scandinavian Journal of Medicine & Science in Sports* 28, no. 2 (February 2018): 717–724. https://doi.org:10.1111/sms.12943

8.   Crum, A. J., and E. J. Langer. "Mind-Set Matters: Exercise and the Placebo Effect." *Psychological Science* 18, no. 2 (2007): 165–171.

9.   Zahrt, O. H., et al. "Effects of Wearable Fitness Trackers and Activity Adequacy Mindsets on Affect, Behavior, and Health: Longitudinal Randomized Controlled Trial." *Journal of Medical Internet Research* 25, e40529 (2023). https://doi.org:10.2196/40529

10.   Kahneman, D., O. Sibony, and C. R. Sunstein. *Noise: A Flaw in Human Judgement.* Little, Brown Spark, 2021.

11.   Langer, E. *The Mindful Body: Thinking Our Way to Chronic Health.* Ballantine Books, 2023.

12.   Stanforth, D., P. R. Stanforth, M. A. Steinhardt, and M. Mackert. "An Investigation of Exercise and the Placebo Effect." *American Journal of Health Behavior* 35, no. 3 (May 2011): 257–268.

13.   Passie, T. "History of the Use of Hallucinogens in Psychiatric Treatment." In *Handbook of Medical Hallucinogens*, edited by C. S. Grob and J. Grigsby, 95. Guilford Press, 2021.

14. Flaherty, E., T. Sturm, and E. Farries. "The Conspiracy of Covid-19 and 5G: Spatial Analysis Fallacies in the Age of Data Democratization." *Social Science & Medicine* 293, (January 2022). https://doi.org:10.1016/j.socscimed.2021.114546

15. Moore, R. "David Icke on Ayahuasca & Having Visions of the Future." YouTube. April 18, 2020. https://www.youtube.com/watch?v=L1oDgUDIcKA

16. Pace, B. "Jake Angeli: The Psychedelic Guru Who Stormed the Capitol." *Psymposia*, January 7, 2021.

17. Kearney, M. D., S. C. Chiang, and P. M. Massey. "The Twitter Origins and Evolution of the COVID-19 'Plandemic' Conspiracy Theory." *Harvard Kennedy School Misinformation Review* 1, (October 2020).

18. Hartogsohn, I. *American Trip: Set, Setting, and the Psychedelic Experience in the Twentieth Century.* MIT Press, 2020; Winkelman, M. J. "The Evolved Psychology of Psychedelic Set and Setting: Inferences Regarding the Roles of Shamanism and Entheogenic Ecopsychology." *Frontiers in Pharmacology* 12, (February 2021); Hartogsohn, I. "Set and Setting in the Santo Daime." *Frontiers in Pharmacology* 12, (May 2021); Lansky, E. S. P. Book Review of Ido Hartogsohn's *American Trip: Set, Setting, and the Psychedelic Experience in the Twentieth Century. Frontiers in Psychology* 12, (2021): 3227; Pontual, A. A. D. D., L. F. Tófoli, C. F. Collares, J. G. Ramaekers, and C. M. Corradi-Webster. "The Setting Questionnaire for the Ayahuasca Experience: Questionnaire Development and Internal Structure." *Frontiers in Psychology* 12, (June 2021); Carhart-Harris, R., in *Synthesis Retreat*. Synthesis, 2020.

19. Grierson, B. *What Makes Olga Run? The Mystery of the Ninety-Something Track Star and What She Can Teach Us About Living Longer, Happier Lives.* Henry Holt, 2014.

20. Raz, A. "Hypnosis: A Twilight Zone of the Top-Down Variety: Few Have Never Heard of Hypnosis but Most Know Little About the Potential of This Mind-Body Regulation Technique for Advancing Science." *Trends in Cognitive Sciences* 15, no. 12 (December 2011): 555–557. https://doi.org/10.1016/j.tics.2011.10.002

21. Machiavelli, N. *Il Principe*. 1532.

22. Shermer, M. "Patternicity: Finding Meaningful Patterns in Meaningless Noise." *Scientific American* 299, no. 6 (December 2008). https://www.scientificamerican.com/article/patternicity-finding-meaningful-patterns

23. Szechtman, H., E. Woody, K. S. Bowers, and C. Nahmias. "Where the Imaginal Appears Real: A Positron Emission Tomography Study of Auditory Hallucinations." *PNAS* 95, no. 4 (February 1998): 1965–1960. https://doi.org/10.1073/pnas.95.4.1956

24. Raz, A., T. Shapiro, J. Fan, and M. I. Posner. "Hypnotic Suggestion and the Modulation of Stroop Interference." *Archives of General Psychiatry* 59, no. 12 (December 2002): 1155–1161.

25. Portela, L. *The Science of Spirit: Parapsychology, Enlightenment and Evolution.* Toplight Books, 2021.

26. Kuhn, G. *Experiencing the Impossible: The Science of Magic.* MIT Press, 2019.

27. Raz, A., I. Kirsch, J. Pollard, and Y. Nitkin-Kaner. "Suggestion Reduces the Stroop Effect." *Psychological Science* 17, no. 2 (February 2006): 91–95. https://doi.org/10.1111/j.1467-9280.2006.01669.x

INDEX